THE AFRICAN AMERICAN FAMILY ALBUM

AMERICAN FAMILY ALBUMS

THE AFRICAN AMERICAN FAMILY ALBUM

DOROTHY AND THOMAS HOOBLER

Introduction by Phylicia Rashad

OXFORD UNIVERSITY PRESS • NEW YORK • OXFORD

Oxford University Press

Oxford New York
Athens Auckland Bangkok Bombay
Calcutta Cape Town Dar es Salaam Delhi
Florence Hong Kong Istanbul Karachi
Kuala Lumpur Madras Madrid Melbourne
Mexico City Nairobi Paris Singapore
Taipei Tokyo Toronto

and associated companies in
Berlin Ibadan

Design: Sandy Kaufman
Layout: Greg Wozney
Consultant: David Thackery, curator of local and family history, Newberry Library, Chicago

Published by Oxford University Press, Inc.,
200 Madison Avenue, New York, New York 10016

Library of Congress Cataloging-in-Publication Data

Hoobler, Dorothy.
The African American family album / Dorothy and Thomas Hoobler; introduction by Phylicia Rashad
p. cm. — (American family albums)
Includes bibliographical references and index.
1. Afro-American families—Juvenile literature. 2. Afro-Americans—History—Juvenile literature.
I. Hoobler, Thomas. II. Title. III. Series.
E185.86.H72 1994

973'.0496073—dc20 94-34697
CIP
AC

ISBN 0-19-508128-5 (lib. ed.); ISBN 0-19-509460-3 (trade ed.); ISBN 0-19-510172-3 (series, lib. ed.)

9 8 7 6 5 4 3 2

Printed in the United States of America
on acid-free paper

Cover: The S. R. Hackett family in 1917

Frontispiece: A California family around 1900

Contents page: Annie R. Lee of San Antonio, Texas

CONTENTS

Phylicia Rashad's talents span the fields of drama, music, and comedy. She has been honored with three NAACP Image Awards and with the People's Choice Award for Best Actress for her role in "The Cosby Show."

She was born Phylicia Ayers-Allen in 1948, into a Houston family that stressed the benefits of education, hard work, family life, and creativity. Her sister, Debbie Allen, is a dancer, choreographer, actress, and singer. Her brother, Tex Allen, is a jazz musician. Their father was Andrew Allen, a prominent Houston dentist. And their mother was Vivian Ayers, a scholar and poet whose book Spice of Dawns *was nominated for the Pulitzer Prize. Rashad credits her parents with urging the children to fulfill their potential: "I was very fortunate to have had parents who wanted to give, could give, and did give. If I didn't get another thing in life, truly, I am blessed with having had that alone."*

Ms. Rashad graduated magna cum laude from Howard University in 1970 with a degree in theater and soon moved to New York, where she managed to work fairly steadily with theater companies and filled in, when necessary, with jobs as a typist. Her big break came when she was offered a part in The Wiz *on Broadway. She has also appeared on Broadway in* Into the Woods *and* Jelly's Last Jam, *among other productions.*

Rashad has also performed as a singer and has acted in various TV shows and series. She is probably best known for her role as Clair Huxtable in "The Cosby Show." Rashad says her real-life experience as a mother helped her land that role. "I realized that I was often acting out a scene on stage that had played out for real *in my living room. I guess the truth of that just comes through."*

In 1986, Phylicia Ayers-Allen married NBC sportscaster Ahmad Rashad after he proposed to her on the air. They currently live in a New York City suburb.

Lloyd Allen, Sr., and Goldia Matilda Allen, Rashad's paternal grandparents, had this portrait taken soon after their wedding in about 1908.

Rashad's great-aunt Fannie Jackson in Lobdell, Louisiana.

The Ayers family in Chester, South Carolina, on Christmas Day 1953. Phylicia Ayers-Allen is seated on the floor, second from left, next to her brother, Tex, wearing a Cub Scouts bandanna. Her sister, Debbie, is seated on the floor, second from right. Her father is in the back row (center), standing behind her mother, who holds a baby cousin.

Rashad with her son, Billy, and her daughter, Condola Phyleia.

Ahmad and Phylicia Rashad jointly present an award in 1986.

INTRODUCTION

by Phylicia Rashad

y father, Dr. Andrew A. Allen, Sr., was born in Lobdell, Louisiana, in 1920. He was the 6th of 10 children born to Lloyd Allen, Sr., and Goldia Jackson Allen. Those were hard times, and "Papa Lloyd," who had become a fireman for the Southern Pacific Railroad, decided to move the family from the farming community in Louisiana to Houston, Texas, to secure a better education for his children. And indeed, my father went on to St. Augustine's College in Raleigh, North Carolina, and then graduated from Howard University Dental College in 1945.

My mother, the poet Vivian Ayers, was born and reared in the Presbyterian community of Chester, South Carolina, where she and her sister attended the private Brainerd Institute. Her parents were both college-trained at a time when only 27 percent of African Americans in South Carolina were receiving any education. Vivian Myrtle Graham attended Hampton Institute and mastered all the crafts of tailoring. My grandfather, Robert Douglas Ayers, led his class at Johnson C. Smith College in math and took over his father's blacksmithing business during the Great Depression. His second marriage, to Bessie Lewis, a teacher with a great love for literature, brought the rigors of poetry and academia full-fledged into the family.

When I was five, we drove from Houston to Chester, about 1,800 miles away. I was so excited. It was not my first visit to South Carolina, but now I was older, understanding and remembering more. It was the holiday season. We left before dawn and made our first rest stop in Lobdell, Louisiana. My great-aunt Fannie Jackson and her younger brother, Ferdie, still lived in the house that had been built by "Papa Lloyd" and my great-grandfather Chanton Jackson. My father loved Aunt Fannie as his own mother, for she had been caretaker to all, mending shirts and broken toys, overseeing homework, and preparing delicious meals for the family. She also had a native knowledge of herbs, wild plants and roots, and their medicinal properties. Every morning, before the sun rose, Fannie and Ferdie were up and about, tending to the many chores on the farm. Animals had to be fed, cows had to be pastured, and they planted and harvested many vegetables. How I loved the many fields and pastures of Lobdell! I was beginning to learn that the good life is synonymous with hard work and continuing education.

When we arrived in South Carolina the following day or so, it was a festive time, and everyone was waiting for us. There were Christmas carols and nativity programs in the church. My most vivid memory was meeting for the first time a tall, slender, handsome man with red-toned skin—who was "Papa Graham," my great-grandfather of Cherokee descent.

Sitting around the hearth and hearing grownups talk, I learned that indeed we had come of many cultures. I was descended of ancestors in Africa, Europe, and America. In my heart I embraced them all. In reflecting on the total experience of growing up, and now as a woman with a family of my own, I feel that we are truly a modern family, having the capacity to accept many cultures with love and respect.

Surely this was a part of my thinking when at the age of 11, I wrote: "Our greatest power is ourselves, because together we are ONE!"

Phylicia Rashad

Two of the magnificent brass heads that were sculpted by the West African people of Ife, in today's Nigeria.

CHAPTER ONE

HOMELAND

I teach kings the history of their ancestors so that the lives of the ancients might serve them as an example, for the world is old, but the future springs from the past." So spoke a 20th-century West African *griot,* or oral historian, whose family preserved the history of their nation's people through memory.

Africa's history stretches far into the past, even beyond the memory of the griots. Modern scientists agree that the ancestor of the first human being was born in Africa, perhaps 200,000 years ago. Five thousand years before the birth of Christ, one of the world's oldest civilizations arose along the Nile River in northeast Africa. Throughout the world's second-largest continent, many distinct groups of people have developed their own customs, literature, religions, science, and art. It would require numerous books to relate the story of all of them.

It was in West Africa, however, that the ancestors of the majority of African Americans were born. Here, at least 2,500 years ago, people living along the Niger River began to plant crops such as sorghum and millet. They also caught fish in the river and herded cattle on the vast grasslands of the region, known as the savanna. The abundance of food created the conditions in which civilization began.

More than 2,000 years ago, artists of the Nok culture in today's northern Nigeria sculpted small terra-cotta heads with very distinctive features. The Nok people also developed iron farm tools. (The prestige of those who made tools was so great that in later times, the kings in some parts of the region were also the chief blacksmiths.)

Over time, West Africans learned to make cloth from the fibers of the cotton plant and to use the indigo plant, berries, and herbs to dye it. Towns sprang up around marketplaces where people brought their crops to trade.

South and west of the fertile Niger River region, other West Africans lived in dense rain forests along the seacoast. Instead of farming, they continued to hunt elephants and other game for their food. But in the earth of their forest homeland they found gold. Prized by people the world over, this gold soon attracted traders to the region.

Until around 2,000 years ago, there was little contact between the West Africans and the rest of the world. A great natural barrier, the Sahara Desert, separated the people of the interior from the Mediterranean seacoast. Around the time of Christ's birth, however, Berber people from North Africa began to use camels to cross the desert. Among the goods they carried was a product that had great appeal for the West Africans—salt, valued not only for its taste but also because it could be used to preserve food.

In what became known as the "silent trade," the Berbers announced their arrival by beating drums. After setting on the ground the goods they had to trade, they withdrew to a safe distance. The West Africans arrived and placed a quantity of gold alongside the salt or cloth that the Berbers had left. They, too, moved out of sight to let the Berbers return. If the Berbers thought the offer was fair, they took the gold and beat their drums. If not, they left the gold on the ground; the African traders could either increase their offer or take the gold away.

The wealth produced by this trade led to the rise of the first great West African empire. Through a series of conquests, a group of people called the Soninke gradually established control over a large region of West Africa, which they called Ghana. (The empire of Ghana should not be confused with the modern nation of the same name, which lies 1,000 miles farther south.) The rulers of Ghana prospered by placing a tax on all gold and other goods that passed through their territory. It was said that one emperor possessed a nugget of gold so large that he could tether his horse to it.

Ghana reached the height of its power around the year 1000. By that time Arabs had swept across North Africa, spreading their religion, Islam. The Arabs expanded the trade with West Africa, and great caravans of camels began to cross the Sahara. Along with trade goods, they brought scholars who sought converts to Islam. The emperors of Ghana welcomed them and allowed mosques to be built where Muslims could worship.

An Arab historian, El-Bakri, wrote in 1067 that the emperor of Ghana "can put 200,000 warriors in the field, more than 40,000 of them armed with bow and arrow." El-Bakri said that the emperor was so wealthy that his horses were "dressed in cloth of gold" and his dogs "wore collars of gold and silver."

About a century later, however, the empire of Ghana was weakened by a series of rebellions. In 1235, it fell to the warriors of the Sosso kingdom, one of the states under Ghana's control. Soumaoro Kanté, the Sosso king, was in turn defeated by Sundiata, a prince of the Mandingo people. Sundiata founded the Mali Empire, which eventually extended from the city of Gao, in today's nation of Mali, to the Atlantic Ocean, 1,300 miles west.

Islamic influence in West Africa grew during the time of the Mali Empire. Sundiata's grandson, Mansa (emperor) Musa, became a devout Muslim. In the year 1324, Mansa Musa set out to fulfill one of his religious duties—to make the *hajj,* or pilgrimage to Mecca, the city in Arabia where Muhammad founded the religion of Islam.

Mansa Musa used the journey to show the world the grandeur of his empire. Accompanied by 60,000 of his subjects and a retinue of bodyguards with gold-tipped spears, Musa brought along 80 camels, each of which carried 300 pounds of gold dust. In every town and city along the way, the emperor awed the inhabitants with his generosity, distributing alms to the poor and bestowing lavish presents on his hosts. The news of his journey spread to Europe, where a map of Africa showed a picture of Musa holding a nugget of gold.

A 19th-century chief of the Zulu people of South Africa with his family.

An Arab writer visited Mali a few years after Musa's death and reported with amazement that he could travel throughout the vast kingdom without fear of robbers. Cities such as Timbuktu and Djenne had mosques, libraries, and schools where Islamic scholars studied and taught.

The religion of Islam, however, did not have as great an effect outside the cities. The farmers and herders in the savanna, and the people along the coasts, retained their traditional customs and religions. Agriculture was a community effort: adults and children all had a role to play. The land itself belonged to everyone; working for the common good was the highest ideal. The Lozi people had a proverb: "Go the way that many people go; if you go alone you will have reason to lament."

The Africans obtained everything they needed from the bounty of the natural world. Parts of the baobab tree, for example, could be used to make bread, a red dye, a spice, and a liquid with healing properties. Living in harmony with nature, the Africans revered its forces. Their proverbs and tales reflect a belief in nature spirits who influenced rainfall, harvests, floods, and other natural phenomena. The music of Africa, played on drums and other percussion and stringed instruments, was often played at ceremonies honoring these spirits. Villagers also built shrines to honor the spirits of their ancestors, who were thought to remain a part of the living world.

Beyond the area controlled by the Mali Empire, smaller kingdoms arose. One example is the kingdom of Ife, in what is today Nigeria. Ife is famous for its sculpture, in particular the wood, ivory, and brass heads that may have been used in funeral ceremonies.

The *oba,* or king, of nearby Benin asked the ruler of Ife to send an artist to teach the secrets of making brass sculpture. The people of Benin developed the art of making brass wall plaques, which make up a kind of history book showing great events of the past.

In the 15th century, the Songhai people, who lived in the eastern part of the Mali Empire, set out on a series of conquests under their ruler, Sonni Ali. He established a new empire around the middle part of the Niger River.

After Sonni Ali's death, one of his generals, Muhammad ben Abu Bekr, seized power. When one of Sonni Ali's daughters accused him of stealing the throne, Muhammad wryly adopted the title Askia ("thief"). Nonetheless, Askia Muhammad's long reign (1493–1528) was a time of prosperity. He created a system of government that made the Songhai Empire stronger than either Mali or Ghana, sending his own officials to govern all parts of his domain.

Askia Muhammad, a good Muslim who made the *hajj* to Mecca, rebuilt the city of Timbuktu, which had been destroyed in the fighting earlier. He founded a university there that became a center of Islamic learning and culture. Leo Africanus, a Moroccan visitor to Timbuktu in 1513, wrote: "Here there are many doctors, judges, priests and other learned men, that are well maintained at the king's cost. Various manuscripts and written books are brought here out of Barbary and sold for more money than any other merchandise."

Warriors, male and female, in the French colony of Dahomey. After 1945, the peoples of Africa began to regain their independence, and Dahomey is today the West African nation of Benin.

Leo Africanus found goods from all parts of the world for sale in the Songhai Empire—Arabian horses, bridles and spurs, fine cloth from Venice and Turkey, European swords, and African slaves. As in other parts of the world, slavery was accepted in Africa. In many ways, however, it differed from the slavery practiced later in America. Many slaves in Africa were prisoners of war taken in battle. Some were later ransomed; those who were not often became part of the group that had captured them, eventually achieving independent status. People could also be enslaved for nonpayment of debt, or as punishment for crimes. Some slaves served as advisers to kings. Most important, there was no idea of a permanent class of people who were slaves by birth.

The development of the trans-Sahara trade routes resulted in an increase in the African slave trade. Muslim Arabs often accepted slaves along with gold in exchange for their goods. However, the prophet Muhammad, the founder of Islam, taught that slaves should be treated well, and that the liberation of a slave was an act of merit.

The Arabs took African slaves with them when they conquered parts of Portugal and Spain. The institution of slavery survived when these countries regained their lost territory in the 15th century. At the same time, Prince Henry the Navigator of Portugal began to search for a sea route to Asia. Henry sent the first European ships down the west coast of Africa, hoping that they could find the southernmost tip of the continent.

When the Portuguese reached the Niger River delta, a fisherman was the first to see them, according to a traditional African story. The fisherman was seized with panic. He rushed back to his village to tell what he had seen. "Whereupon he and the rest of the town set out to purify themselves—that is, to rid themselves of the influence of the strange and monstrous thing that had intruded into their world."

Especially in rural areas, many Africans retain their traditional customs today. The coiled necklaces worn by this woman in Uganda indicate that she is married.

For centuries, the people who live along the Niger River of West Africa have caught fish as part of their diet. The great empires of Ghana, Mali, and Songhai all arose along the Niger.

LIFE IN AFRICA

One of the first enslaved Africans who set down memories of his homeland wrote under the name Venture. In 1737, when he was about eight years old, Venture was captured and brought to the United States. He wrote his life story in 1798, when he was nearly 70 years old.

I was born at Dukandarra, in Guinea, about the year 1729. My father's name was Saungm Furro, Prince of the tribe of Dukandarra. My father had three wives. Polygamy was not uncommon in that country, especially among the rich, as every man was allowed to keep as many wives as he could maintain. By his first wife he had three children. The eldest of them was myself, named by my father, Broteer.... I descended from a very large, tall and stout race of beings, much larger than the generality of people in other parts of the globe, being commonly considerable above six feet in height, and every way well proportioned.

The first thing worthy of notice which I remember, was a contention between my father and mother, on account of my father marrying his third wife without the consent of his first and eldest, which was contrary to the custom generally observed among my countrymen. In consequence...my mother left her husband and country, and travelled away with her three children to the eastward. I was then five years old. She took not the least sustenance with her, to support either herself or children. I was able to travel along by her side; the other two of her offspring she carried, one on her back, the other, being a sucking child, in her arms. When we became hungry, our mother used to set us down on the ground and gather some of the fruits that grew spontaneously in that climate....

Thus we went on our journey until the second day after our departure from Dukandarra, when we came to the entrance of a great desert. During our travel in that, we were often affrighted with the doleful howlings and yellings of wolves, lions and other animals. After five days' travel we came to the end of this desert, and immediately entered into a beautiful and extensive...country. Here my mother was pleased to stop and seek a refuge for me. She left me at the house of a very rich farmer....

My new guardian...put me into the business of tending sheep immediately after I was left with him. The flock, which I kept with the assistance of a boy, consisted of about forty. We drove them every morning between two and three miles to pasture, into the wide and delightful plains. When night drew on, we drove them home and secured them in the cote....

A large river runs through this country in a westerly course. The land for a great way on each side is flat and level, hedged in by a considerable rise in the country at a great distance from it. It scarce ever rains there, yet the land is fertile; great dews fall in the night which refresh the soil. About the latter end of June or first of July, the river begins to rise, and gradually increases until it has inundated the country...to the height of seven or eight feet. This brings on a slime which enriches the land surprisingly. When the river has subsided, the natives begin to sow and plant, and the vegetation is exceeding rapid. [Venture's parents settled their quarrel and his father sent a man with a horse to bring him home.]

Around 1755, an 11-year-old African named Olaudah Equiano was kidnapped and sold into slavery. Years later he bought his freedom and wrote an autobiography. Equiano began by describing the African empire where he was born.

That part of Africa, known by the name of Guinea, to which the trade for slaves is carried on, extends along the coast above [more than] 3400 miles, from Senegal to Angola, and includes a variety of kingdoms. Of these the most considerable is the kingdom of Benin, both as to extent and wealth, the richness and cultivation of the soil, the power of its king, and the number and warlike disposition of its inhabitants.... This kingdom is divided into many provinces or districts, in one of the most remote and fertile of which I was born, in the year 1745, situated in a charming, fruitful vale named Essaka. The distance of this province from the capital of Benin and the sea coast must be very considerable, for I had never heard of white men or Europeans, nor of the sea; and our subjection to the king of Benin was little more than nominal.... My father was one of those elders or chiefs [who governed the local area], and was styled Embrenche, a term, as I remember, importing the highest distinction, and signifying in our language a *mark* of grandeur. This mark is conferred on the person entitled to it, by cutting the skin across at the top of the forehead, and drawing it down to the eyebrows.... I had seen it conferred on one of my brothers, and I also was *destined* to receive it by my parents. Those Embrenche, or chief men, decided disputes and punished crimes, for which purpose they always assembled together....

We are almost a nation of dancers, musicians, and poets. Thus every great event, such as a triumphant return from battle or other cause of public rejoicing, is celebrated in public dances, which are accompanied with songs and music suited to the occasion.... We have many musical instruments, particularly drums of different kinds, a piece...which resembles a guitar, and another much like a stickado [xylophone]. These last are chiefly used by betrothed virgins, who play on them on all grand festivals.

As our manners are simple, our luxuries are few. The dress of both sexes...generally consists of a long piece of calico, or

As Others Saw Them

One of the earliest European visitors to Benin, in 1589, described the variety of goods available in the marketplace:

Pepper and elephant teeth, oil of palm, cloth made of cotton wool very curiously woven, and cloth made of the bark of palm trees were procured in exchange for cloth, both linen and woolen, iron works of sundry sorts, Manillos or bracelets of copper, glass beads and coral.... They have good store of soap...also many pretty fine mats and baskets that they make, and spoons of elephant's teeth very curiously wrought with divers proportions of fowls and beasts made upon them.

The title page from Olaudah Equiano's autobiography. He adopted the name Gustavus Vassa in England after gaining his freedom.

muslin, wrapped loosely round the body.... This is usually dyed blue, which is our favorite color. It is extracted from a berry, and is brighter and richer than any I have seen in Europe. Besides this, our women of distinction wear golden ornaments, which they dispose...on their arms and legs....

Bullocks, goats, and poultry supply the greatest part of their food. These constitute likewise the principal wealth of the country, and the chief articles of its commerce. The flesh is usually stewed in a pan; to make it savory we sometimes use pepper, and other spices, and we have salt made of wood ashes. Our vegetables are mostly plantains, eadas [an edible root], yams, beans, and Indian corn. The head of the family usually eats alone; his wives and slaves have also their separate tables. Before we taste food we always wash our hands; indeed, our cleanliness on all occasions is extreme.... After washing, libation is made, by pouring out a small portion of the drink on the floor, and tossing a small quantity of the food in a certain place, for the spirits of departed relations, which the natives suppose to preside over their conduct and guard them from evil....

We have fire-arms, bows and arrows, broad two-edged swords and javelins; we have shields also which cover a man from head to foot. All are taught the use of these weapons; even the women are warriors, and march boldly out to fight along with the men.... I was once a witness to a battle in our common [the fields where crops were grown]. We had been all at work in it one day as usual, when our people were suddenly attacked. I climbed a tree...from which I beheld the fight. There were many women as well as men on both sides; among others my mother was there, and armed with a broad sword. After fighting for a considerable time with great fury, and many had been killed, our people obtained the victory, and took their enemy's Chief a prisoner.... Those prisoners which were not sold or redeemed, we kept as slaves; but how different was their condition from that of the slaves in the West Indies! With us, they do no more work than other members of the community, even their master; their food, clothing, and lodging were nearly the same...(except that they were not permitted to eat with those who were free-born).... Some of these slaves even have slaves under them as their own property, and for their own use.

A market day in a West African town. Farmers from the surrounding region bring their crops to sell or exchange for cloth and other manufactured goods.

As to religion, the natives believe that there is one Creator of all things, and that he lives in the sun.... According to some, he smokes a pipe, which is our own favorite luxury. They believe he governs events, especially our deaths or captivity; but as for the doctrine of eternity, I do not remember to have ever heard of it.... [Offerings were made at the graves of dear friends or relations.] When [my mother] went to make these oblations at her mother's tomb, which was a kind of small solitary thatched house, I sometimes attended her. There she...spent most of the night in cries and lamentations. I have been often extremely terrified on these occasions. The loneliness of the place, the darkness of the night, and the ceremony of libation, naturally awful and gloomy, were heightened by my mother's lamentations; and these concurring with the doleful cries of birds, by which these places were frequented, gave an inexpressible terror to the scene.

We compute the year, from the day on which the sun crosses the line [the equator], and on its setting that evening, there is a general shout throughout the land.... The people at the same time make a great noise with rattles...and hold up their hands to heaven for a blessing....

My father, besides many slaves, had a numerous family, of which seven lived to grow up, including myself and my sister, who was the only daughter. As I was the youngest of the sons, I became, of course, the greatest favorite with my mother, and was always with her; and she used to take particular pains to form my mind. I was trained up from my earliest years in the art of war: my daily exercise was shooting and throwing javelins, and my mother adorned me with emblems, after the manner of our greatest warriors. In this way I grew up till I had turned the age of eleven, when an end was put to my happiness [when I was kidnapped into slavery].

A small number of Africans freely immigrated to the United States, and continue to arrive today. Their life stories reveal much about the culture of their homeland. In 1950, Michael Olatunji came to the United States. His homeland was today's Nigeria, then a British colony. Olatunji, one of the world's greatest practitioners of the art of drumming, recalled his youth in West Africa.

I came from a beautiful fishing village called Ajido, about forty-five miles from the capital, Lagos, known in the tra ditional way as Ako. It was a blessing that I grew up there in this kind of place where the tradition was actually taught to young people from the time of birth.... My childhood days I can remember most vividly with satisfaction that even though I didn't have the modern toys that children today play with, yet I played with all of the things that nature has to produce or that nature has blessed mankind with. I lived near the lagoon, I lived near the ocean, so I learn how to fish and to paddle. I followed my brothers and cousins to the farm so I learn how to plant the seed that...later germinate into beautiful flowers, things to eat. I grew up in a situation whereby I enjoy listening to beautiful singing birds early in the morning where the cock crows at five, where the, what most people might consider the noisy waves of the sea, probably remind me today of the thundering rhythms of the drum. I grew up in a situation whereby our music and dance really covers all the [situations] of life because when a child is born there is singing and dancing. When you become an adolescent and you come of age, there is singing and dancing. When you took the giant step of getting married, there is singing and dancing and rejoicing. And we also learned that when a person dies there is singing and rejoicing, drumming. So it became a way of life from childhood.

Langston Hughes (1902–67) was an African American poet who was deeply interested in his African roots. He wrote:

So long,
So far away
Is Africa.
Not even memories alive
Save those that history books create,
Save those that songs
Beat back into the blood—
Beat out of blood with words sad-sung
In strange un-Negro tongue—
So long,
So far away
Is Africa.

Subdued and time-lost
Are the drums—and yet
Through some vast mist of race
There comes this song
I do not understand,
This song of atavistic land,
Of bitter yearning lost
And touching each other natural as dew
In that dawn of music when I
Get to be a composer
And write about daybreak
In Alabama.

Michael Olatunji's concerts in the United States brought the complex and beautiful rhythms of African music to a wider audience.

A GRIOT SPEAKS

While researching the history of his family, the African American writer Alex Haley (at left) went to the nation of Gambia. There, he met a griot who gave him some of the information that appeared in Haley's book Roots.

The griots, *or oral historians, of West Africa memorized the history and great deeds of their people. Mamoudou Kouyaté, a* griot *in the modern nation of Guinea, declared that this method of recording history was superior to writing. He said, "Other peoples use writing to record the past, but this invention has killed the faculty of memory among them. They do not feel the past any more, for writing lacks the warmth of the human spirit."*

Kouyaté tells the story of Sundiata, the greatest of the kings of Mali, who reigned around 700 years ago.

Listen then, sons of Mali, children of the black people, listen to my word, for I am going to tell you of Sundiata, the father of the Bright Country, of the savanna land, the ancestor of those who draw the bow, the master of a hundred vanquished kings.... Maghan Kon Fatta, the father of Sundiata, was renowned for his beauty in every land; but he was also a good king loved by all the people.

[Maghan Kon Fatta had three wives and six children; Sundiata was born to Sogolon Kedjou, the second wife. Kouyaté describes the day of his birth.] The king commanded the nine greatest midwives of Mali to come to Niani, and they were now constantly in attendance on the damsel of Do. The king was in the midst of his courtiers one day when someone came to announce to him that Sogolon's labors were beginning.... One would have thought that this was the first time that he had become a father, he was so worried and agitated. The whole palace kept complete silence.... Suddenly the sky darkened and great clouds coming from the east hid the sun, although it was still the dry season.... A flash of lighting accompanied by a dull rattle of thunder burst out of the east and lit up the whole sky as far as the west. Then the rain stopped and the sun appeared and it was at this very moment that a midwife came out of Sogolon's house, ran to the antechamber and announced to...Maghan that he was the father of a boy....

[The boy was given the name Sogolon Djata, a name that was later shortened to Sundiata. It soon became obvious that the boy had difficulty learning to walk. That was part of his destiny, as Kouyaté relates.] God has mysteries which none can fathom. You, perhaps, will be a king. You can do nothing about it. You on the other hand, will be unlucky, but you can do nothing about that either. Each man finds his way already marked out for him and he can change nothing of it.

[When Djata was seven, his father died. The old king's first wife put her son Dankaran Touman on the throne. One day the first wife humiliated Djata's mother by giving her two baobab leaves, telling her:] "As for me, my son knew how to walk at seven and it was he who went and picked these baobab leaves. Take them then, since your son is unequal to mine." Then she laughed derisively with that fierce laughter which cuts through your flesh and penetrates right to the bone.

[Finding his mother crying, Djata learned of the insult and made up his mind to walk. The royal blacksmith had made a large iron bar. Djata asked that it be brought to his mother's house.] He crept on all-fours and came to the iron bar. Supporting himself on his knees and one hand, with the other hand he picked up the iron bar without any effort.... A deathly silence had gripped all those present. Sogolon Djata closed his eyes, held tight, the muscles in his arms tensed. With a violent jerk he threw his weight on to it and his knees left the ground.... Djata was sweating and the sweat ran from his brow. In a great effort he straightened up and was on his feet at one go—but the great bar of iron was twisted and had taken the form of a bow! [Djata walked to a baobab tree, tore it up by the roots, and brought it to his mother.]

[The first wife now began to fear Djata, and she asked a sorceress to kill him. Djata's mother took her son and fled into exile, where he became known as Sundiata. As Sundiata grew to manhood, Soumaoro Kanté, the blacksmith-king of Sosso, began a series of conquests. He forced Sundiata's half-brother Dankaran Touman to make Mali a tributary state of Sosso. Hearing of this, Sundiata gathered warriors who rode on horseback to meet the Sosso army. Kouyaté describes the battle, in which the Sosso soldiers are called "smiths" and are commanded by Sosso Balla, the son of Soumaoro Kanté:]

The lightning that flashes across the sky is slower, the thunderbolts less frightening and floodwaters less surprising than Sundiata swooping down on Sosso Balla and his smiths. In a trice, Sundiata was in the middle of the Sossos like a lion in the sheepfold. The Sossos, trampled under the hooves of his fiery charger, cried out. When he turned to the right the smiths of Soumaoro fell in their tens, and when he turned to the left his sword made heads fall as when someone shakes a tree of ripe fruit.... Charging ever forward, Sundiata looked for Sosso Balla; he caught sight of him and like a lion bounded towards the son of Soumaoro, his sword held aloft. His arm came sweeping down but at that moment a Sosso warrior came between Djata and Sosso Balla and was sliced like a calabash [gourd]. Sosso Balla did not wait and disappeared from amidst his smiths. Seeing their chief in flight, the Sossos gave way and fell into a terrible rout.

[The other kings of the region submitted to Sundiata as well. This was the beginning of the Mali Empire.] In their newfound peace the villages knew prosperity again, for with Sundiata happiness had come into everyone's home. Vast fields of millet, rice, cotton, indigo...surrounded the villages. Whoever worked always had something to live on. Each year long caravans carried the taxes in kind to Niani.... Sundiata... had made the capital of an empire out of his father's village and Niani became the navel of the earth.

Jalele sainou ane na sainou guissetil dara, tey mague dieki thy soufe guissa yope. (The child looks everywhere and often sees nothing; but the old man, sitting on the ground, sees everything.)

Jama sa bope mo guenne kou la ko waja. (Know thyself better than he does who speaks of thee.)

—sayings of the Wolof people of West Africa

A French missionary priest took this photograph of an African held in a net, destined for slavery. From the 15th to the 19th century, millions of other Africans shared his fate.

CHAPTER TWO

THE MIDDLE PASSAGE

The first slaves from Africa arrived in Portugal in 1442. Over the next 450 years, it is estimated that more than 11 million Africans (some think as many as 40 million) were taken from their homeland—the largest forcible movement of people in history. The majority were shipped to the New World, where Spain and Portugal began to establish colonies in the 1500s. By 1540 more than 10,000 Africans were being taken to the West Indies each year.

The enormous demand for slave laborers in the New World caused the slave trade in Africa to expand far beyond what it had been earlier. Europeans encouraged it by offering wine, guns, and other manufactured products from Europe. To obtain these goods, Africans of the coastal regions raided their neighbors for the sole purpose of obtaining slaves to sell.

Chained together, the slaves were marched to the seacoast, where they were imprisoned in *barracoons,* or holding pens. One of these *barracoons* has been preserved as a museum in the modern nation of Senegal. Leg chains fastened to the stone walls of underground rooms bear testimony to the terror slaves felt. A door that opened onto a rocky cliff over the sea was the exit for slaves who were too unruly, weak, or ill to be sold.

After being sold to a European slave trader, the Africans were branded and transported to a ship anchored offshore. An English ship captain taking slaves to the Barbados Islands remarked, "The negroes are so [unwilling] to leave their own country, that they have often leap'd...into the sea...till they were drowned.... They [have] a more dreadful apprehension of Barbadoes than we can have of hell."

The reality was indeed a kind of hell. Once aboard ship, the enslaved Africans were subjected to the unbelievable cruelty of the "middle passage" to America.

The more slaves that could be crammed aboard, the greater the profits of the voyage would be. Stripped of their clothing, the slaves were chained side by side and forced to lie on the hard wood of the hold. A British slave ship named *Brookes,* 25 feet wide and 100 feet long, carried 609 slaves on one voyage. This was regarded as "tight packing"; some captains preferred "loose packing," for that way not so many slaves died on the voyage.

Countless numbers of them did die. Ventilation was poor, and the heat below deck was so great that one crewman saw "steam coming through the gratings, like a furnace." Bodily wastes were left on the floor. In such conditions, disease was widespread. Some ships carried doctors to keep the valuable cargo alive, but one doctor said that after being below for 15 minutes, "I was so overcome with the heat, stench and foul air that I nearly fainted." By some estimates, one out of every four slaves did not survive the voyage; sharks ate the bodies of the dead that were tossed overboard.

Twice a day the slaves were brought on deck to receive food and water. At mealtimes, some captains forced the slaves to dance and sing because exercise was thought to prevent disease. However, whenever the slaves were allowed out of the hold, they were likely to try to throw themselves over the side. Drowning or sharks were preferable to the horrors below deck, and the slaves believed that their spirits would return to Africa.

Finally, the ghastly voyage—lasting from three weeks to three months—ended. The survivors were led off the ship into a strange new land, destined for a lifetime of slavery. This was the historical beginning of the American experience for the majority of today's African Americans.

As Others Saw Them

After the captured Africans were brought to the shore, they were held in barracoons, *or slave pens, before their brutal voyage. Jean Barbot, a French agent in West Africa, described the slave pens of the late 17th century:*

As the slaves come down to Ouidah from the inland country they are put into a booth or prison, built for that purpose near the beach, all of them together; and when the Europeans are to receive them, they are brought out into a large plain, where the ships' surgeons examine every part of every one of them, to the smallest member, men and women being all stark naked. Such are allowed [judged] good and sound are set on one side...; [each] is marked on the breast with a red-hot iron, imprinting the mark of the French, English, or Dutch companies so that each nation may distinguish their own property, and so as to prevent their being changed by the sellers for others that are worse.... In this particular, care is taken that the women, as the tenderest, are not burnt too hard.

A slave coffle, in which people were chained together to keep them from escaping.

CAPTURED

The African known as Venture described his capture and enslavement in 1737, when he was eight. His father, Saungm Furro, the chief of Dukandarra, received a message from a neighboring people.

That place had been invaded by a numerous army, from a nation not far distant, furnished with...all kinds of arms then in use; that they were instigated by some white nation who equipped and sent them to subdue and possess the country....

[The invading army soon arrived in the territory of Venture's father. Saungm Furro fled with his wife and children, but the enemy discovered them.] My father...immediately began to discharge arrows at them.... It alarmed both me and the women, who...immediately betook ourselves to the tall, thick reeds not far off, and left the old king to fight alone. For some time I beheld him from the reeds defending himself with great courage and firmness, till at last he was obliged to surrender himself into their hands.

They then came to us in the reeds, and the very first salute I had from them was a violent blow on the head with the fore part of a gun, and at the same time a grasp around the neck. I then had a rope put about my neck, as had all the women in the thicket with me, and were immediately led to my father, who was likewise pinioned and haltered for leading.... My father was closely interrogated respecting his money, which they knew he must have. But as he gave them no account of it, he was instantly cut and pounded on his body with great inhumanity.... I saw him while he was thus tortured to death. The shocking scene is to this day fresh in my memory, and I have often been overcome while thinking on it.... He was a man of remarkable strength and resolution, affable, kind and gentle, ruling with equity and moderation.

The army of the enemy...immediately marched towards the sea, lying to the west, taking with them myself and the women....

All of us were then put into the castle and kept for market. On a certain time, I and other prisoners were put on board a canoe, under our master, and rowed away to a vessel belonging to Rhode Island.... I was bought on board by one Robertson Mumford, steward of said vessel, for four gallons of rum and a piece of calico, and called VENTURE, on account of his having purchased me with his own private venture. Thus I came by my name.

Olaudah Equiano described how his happy African childhood ended when he was kidnapped in 1755.

One day, when all [the elders] were gone out to their works and only I and my dear sister were left to mind the house, two men and a woman got over our walls, and in a moment seized us both, and without giving us time to cry out, or make resistance, they stopped our mouths, and ran off with us into the wood. Here they tied our hands, and continued to carry us as far as they could, till night came on, when we reached a small house where the robbers...spent the night.... The next morning we...continued travelling all the day.... [Equiano saw some friends and] began to cry out for their assistance; but my cries had no other effect than to make [the kidnappers] tie me faster and stop my mouth, and then they put me into a large sack.... The next day...my sister and I were then separated.... It was in vain that we besought them not to part us; she was torn from me, and immediately carried away.... I cried and grieved continually; and for several days did not eat anything but what they forced into my mouth.

Asa-Asa was born in Sierra Leone (Asa-Asa called it Bycla) in the early 19th century. In the 1820s, when Asa-Asa was about 13 years old, he was captured and put aboard a French slave ship on the African coast. Fortunately, a storm sent the ship off course to England, where slavery had been abolished in 1772. An English judge freed Asa-Asa and the other slaves. Asa-Asa stayed in England, where he became a Christian. His account of his early life appeared in a book published in 1831.

My father's name was Clashoquin; mine is Asa-Asa. He lived in a country called Bycla, near Egie, a large town.... A great many people, whom we called Adinyes, set fire to Egie in the morning before daybreak; there were some thousands of them. They killed a great many, and burnt all their houses. They stayed two days, and then carried away all the people whom they did not kill....

We all ran away...to the woods...and staid there about four days and nights.... I ran up into a tree; they followed me and brought me down. They tied my feet. I do not know if they found my father and mother, and brothers and sisters: they had run faster than me, and were half a mile farther when I got up into the tree: I have never seen them since....

They carried away about twenty besides me. They carried us to the sea. They did not beat us: they only killed one man, who was very ill and too weak to carry his load; they made all of us carry chickens and meat for our food; but this poor man could not carry his load, and they ran him through the body with a sword. He was a neighbour of ours. When we got to the sea they sold all of us, but not to the same person. They sold us for money; and I was sold six times over, sometimes for money, sometimes for cloth, and sometimes for a gun. I was about thirteen years old. It was about half a year from the time I was taken, before I saw the white people.

Phillis Wheatley

In 1761, when Phillis Wheatley was a child in Africa, she was kidnapped and taken to Boston. Her Quaker master and mistress taught her to read and write. Wheatley started writing poetry soon afterward. In 1773, when she was 20, a book of her poems was published. Though Wheatley was not the first African American poet, she is the best known of the early writers.

During Phillis Wheatley's lifetime, the American colonies fought for independence from Britain—a cause that she supported. She addressed the following poem to the Earl of Dartmouth, who opposed the colonists' demands for greater rights. Wheatley links the colonists' struggle with her own experience:

Should you, my lord, while you pursue my song,
Wonder from whence my love of *Freedom* sprung,
Whence flow these wishes for the common good,
By feeling hearts alone best understood,
I, young in life, by seeming cruel fate
Was snatch'd from *Afric's* fancy'd happy seat:
What pangs excruciating must molest,
What sorrows labour in my parents' breast?
Steel'd was the soul and by no misery mov'd
That from a father seiz'd his babe belov'd
Such, such my case. And can I then but pray
Others may never feel tyrannic sway?

TO BE SOLD on board the Ship *Bance-Iſland*, on tueſday the 6th of *May* next, at *Aſhley-Ferry*; a choice cargo of about 250 fine healthy

NEGROES,

juſt arrived from the Windward & Rice Coaſt. —The utmoſt care has already been taken, and ſhall be continued, to keep them free from the leaſt danger of being infected with the SMALL-POX, no boat having been on board, and all other communication with people from *Charles-Town* prevented.

Auſtin, Laurens, & Appleby.

N. B. Full one Half of the above Negroes have had the SMALL-POX in their own Country.

An advertisement for a slave auction proclaims that the slaves were free from smallpox, one of the diseases that thrived in the crowded slave ships.

This drawing shows how it was possible to pack a cargo of slaves into a space that was only three feet, three inches high.

A VOYAGE OF HORRORS

Olaudah Equiano recalled the day when his captors brought him to the seacoast to be sold.

The first object which saluted my eyes when I arrived on the coast was the sea, and a slave ship, which was then riding at anchor and waiting for its cargo. These filled me with astonishment, which was soon converted into terror, when I was carried on board. I was immediately handled, and tossed up to see if I were sound, by some of the crew; and I was now persuaded that I had gotten into a world of bad spirits and that they were going to kill me. Their [white] complexions, too, differing so much from ours, their long hair, and the language they spoke (which was very different from any I had ever heard), united to confirm me in this belief. Indeed, such were the horrors...at that moment, that, if ten thousand worlds had been my own, I would have freely parted with them all to have exchanged my condition with that of the meanest slave in my own country. When I looked round the ship too, and saw a...multitude of black people of every description chained together, every one of their countenances expressing dejection and sorrow, I no longer doubted of my fate; and quite overpowered with horror and anguish, I fell motionless on the deck and fainted. When I recovered a little, I found some black people about me, who I believed were some of those who had brought me on board.... I asked them if we were not to be eaten by those white men with horrible looks, red faces, and long hair. They told me I was not, and one of the crew brought me a small portion of spiritous liquor in a wine glass but, being afraid of him, I would not take it.... Soon after this, the blacks who brought me on board went off, and left me abandoned to despair....

I was soon put down under the decks, and there I received such a salutation in my nostrils as I had never experienced in my life: so that with the loathsomeness of the stench and crying together, I became so sick and low that I was not able to eat.... I now wished for the last friend, death, to relieve me; but soon, to my grief, two of the white men offered me eatables, and on my refusing to eat, one of them held me fast by the hands...and tied my feet while the other flogged me severely. I had never experienced anything of this kind before, and although, not being used to the water, I naturally feared that element the first time I saw it, yet nevertheless could I have got over the nettings I would have jumped over the side.

Though the United States banned the further importation of slaves in 1808, smugglers continued to practice the lucrative trade. This engraving from a newspaper in 1860 depicts a group of slaves on a ship captured by the U.S. Navy. Attempts were made to return such people to Africa.

A chart used by slave ship captains shows an efficient way of loading a human cargo into the smallest possible space.

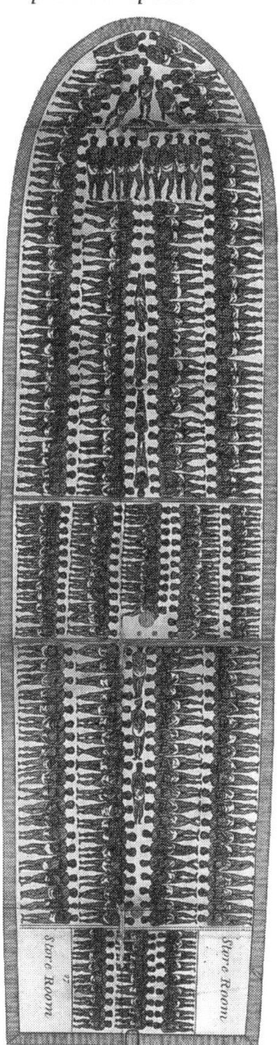

When the ship carrying Equiano sailed, the slaves were chained together below deck.

The air soon became unfit...from a variety of loathsome smells, and brought on a sickness among the slaves, of which many died.... The shrieks of the women, and the groans of the dying, rendered the whole a scene of horror almost inconceivable. Happily perhaps for myself, I was soon reduced so low here that it was thought necessary to keep me almost always on deck; and from my extreme youth I was not put in fetters.... Every circumstance I met with...heightened my apprehensions, and my opinion of the cruelty of the whites.

One day they had taken a number of fishes; and when they had killed and satisfied themselves with as many as they thought fit, to our astonishment who were on deck, rather than give any of them to us to eat, as we expected, they tossed the remaining fish into the sea again, although we begged and prayed for some as well as we could, but in vain.... One day, when we had a smooth sea and moderate wind, two of my wearied countrymen who were chained together...preferring death to such a life of misery, somehow made through the nettings and jumped into the sea; immediately another quite dejected fellow, who, on account of his illness, was suffered to be out of irons, also followed their example; and I believe

In 1854, Theodore Canot, a slave trader who made many voyages between Africa and America, published an account of his experiences. Canot writes chillingly of the practical matters of transporting the slaves:

An African factor [trader] of fair repute is ever careful to select his human cargo with consummate prudence, so as not only to supply his employers with athletic labourers, but to avoid any taint of disease that may affect the slaves in their transit to Cuba or the American main[land]. Two days before embarkation, the head of every male and female is neatly shaved; and if the cargo belongs to several owners, each man's *brand* is impressed on the body of his respective negro. This operation is performed with pieces of silver wire, or small irons fashioned into the merchant's initials, heated just hot enough to blister without burning the skin....

On the appointed day, the barracoon or slave-pen is made joyous by the abundant "feed" which signalises the negro's last hours in his native country. The feast over, they are taken alongside the vessel in canoes; and as they touch the deck, they are entirely stripped, so that women as well as men go out of Africa as they came into it—naked. This precaution, it will be understood, is indispensable; for perfect nudity, during the whole voyage, is the only means of securing cleanliness and health....

At sundown, the process of stowing the slaves for the night is begun. The second mate and boatswain descend into the hold, whip in hand, and range the slaves in their regular places.... each negro lies on his right side, which is considered preferable for the action of the heart. In allotting places, particular attention is paid to size, the taller being selected for the [middle] of the vessel, while the shorter and younger are lodged near the bows....

It is very probable that many of my readers will consider it barbarous to make slaves lie down naked upon a board, but let me inform them that native Africans are not familiar with the use of feather-beds, nor do any but the free and rich in their mother country indulge in the luxury even of a mat or raw-hide.... I am of opinion, therefore, that emigrant slaves experience very slight inconvenience in lying down on the deck.

many more would very soon have done the same if they had not been prevented by the ship's crew who were instantly alarmed.... Two of the wretches were drowned, but they got the other, and afterwards flogged him unmercifully, for thus attempting to prefer death to slavery.

After Asa-Asa was freed in England, he wrote a memoir in which he recalled his first days as a slave.

We were taken in a boat from place to place, and sold at every place we stopped at. In about six months we got to a ship, in which we first saw white people: they were French. They bought us. We found here a great many other slaves; there were about eighty, including women and children. The Frenchmen sent away all but five of us into another very large ship. We five stayed on board till we got to England, which was about five or six months. The slaves we saw on board the ship were chained together by the legs below deck, so close they could not move. They were flogged very cruelly: I saw one of them flogged till he died; we could not tell what for. They gave them enough to eat. The place they were confined in below deck was so hot and nasty I could not bear to be in it. A great many of the slaves were ill, but they were not attended to. They used to flog me very bad on board the ship: the captain cut my head very bad one time....

I should like much to see my friends again, but I do not now wish to go back to them; for if I go back to my own country, I might be taken as a slave again. I would rather stay here, where I am free, than go back to my country to be sold.

Olaudah Equiano described the arrival of the slave ship at the island of Barbados, an English colony in the West Indies. (Later, Equiano would go to Virginia.)

The merchants and planters now came on board.... They...examined us attentively. They also made us jump, and pointed to the land, signifying we were to go there. We thought by this, we should be eaten by these ugly

A book written by a slave smuggler in 1856 shows the author distributing food and water in the hold of his ship.

men, as they appeared to us; and, when soon after we were all put down under the deck again, there was dread and trembling among us, and nothing but bitter cries to be heard all the night.... At last the white people got some old slaves from the land to pacify us. They told us we were not to be eaten, but to work, and were soon to go on land, where we should see many of our country people. This report eased us much. And sure enough, soon after we were landed, there came to us Africans of all languages.

We were conducted immediately to the merchant's yard, where we were all pent up together, like so many sheep in a fold, without regard to sex or age....

We were not many days in the merchant's custody, before we were sold after their usual manner, which is this: On a signal given (as the beat of a drum), the buyers rush at once into the yard where the slaves are confined, and make choice of that parcel they like best. The noise and clamor with which this is attended and the eagerness visible in the countenances of the buyers, served not a little to increase the apprehension of terrified Africans.... In this manner...are relations and friends separated, most of them never to see each other again. I remember, in the vessel in which I was to be brought over...there were several brothers, who, in the sale, were sold in different lots; and it was very moving on this occasion, to see and hear their cries at parting.

In 1992, in an annual ceremony in Brooklyn, New York, African Americans mourn slaves who died crossing the Atlantic Ocean on their way to America's shores.

A slave couple in front of a log cabin that they themselves might have built. Their faces show the effects of a lifetime of toil serving their master.

CHAPTER THREE

SLAVERY

herever the Spanish explorers went in the New World, Africans accompanied them. In 1513, Africans marched through the jungles of Panama with Vasco Nuñez de Balboa and beheld the vast waters of the Pacific. Africans took part in the 16th-century Spanish conquests of the Native American empires of Mexico and Peru. Africans stepped ashore in Florida in 1519 with Juan Ponce de Leon. Forty-six years later, African slave laborers built what is today the oldest city in the United States, St. Augustine, Florida.

Seldom were the names of any of these Africans preserved in history, for they were slaves. The first African in the United States—whose name we know—was Estevanico ("little Stephen"). His fabulous career as an explorer took him from Florida to Texas and later into New Mexico.

The first Africans to settle in what is now the United States arrived in 1526 with a Spanish colonizing party near the Pee Dee River in today's South Carolina. The colony failed when some of the slaves rebelled and escaped. They hoped to find safety among the Native Americans in the region—and perhaps they did.

In 1619, a Dutch ship brought 10 Africans to Jamestown, Virginia—the first to arrive in an English colony in North America. These Africans may have been treated as indentured servants rather than slaves. Indentured servants signed contracts requiring them to serve for a set period of years in return for their passage to America. Indeed, about half of the early *white* settlers in British North America arrived as indentured servants. Because white indentured servants provided the main source of labor, the institution of slavery took hold slowly. By 1660 there were only about 300 Africans in Virginia, and smaller numbers in New England. Some had won their freedom, and a few even owned their own plantations.

There were also African slaves in the Dutch colony of New Amsterdam (today's New York City), but they had many legal protections, including the right to farm their own land and to testify against whites in court. In 1664, however, the British captured New Amsterdam, renaming it after the king's brother, the Duke of York. Under British rule, "Negro servants" in New York were ordered to be held in perpetual slavery—the first time this had occurred in the British American colonies.

There was a reason for this new policy. A year earlier, the Duke of York had become a partner in an English company that intended to seize control of the lucrative African slave trade. From that point on, encouraged by the British government, more and more slaves came directly from Africa to North America. The importation of slaves reached its peak between 1741 and 1760, when more than 5,000 per year arrived in the British North American colonies.

Particularly in the South, the influx of slaves created new opportunities for profitable forms of agriculture. In Virginia, tobacco, planted and harvested by slaves, became the major export. Rice and indigo, introduced by African slaves from their homeland, became the mainstay of colonial plantations in the Carolinas.

The increase in the African population caused the colonial legislatures to pass laws restricting their rights. In 1691 Virginia outlawed the practice of liberating slaves. Free Africans, who were previously allowed to vote, were denied this right in Virginia in 1723. Other colonies eventually followed suit. Though slavery was forbidden by Georgia's charter when that colony was founded in 1733, its legislature reversed the policy 17 years later.

Fewer numbers of African slaves came to the northern colonies, partly because large-scale farming did not catch on there. Northerners prospered, however, from the participation of their shipping companies in the slave trade.

Most of the English colonists were staunch Protestants, and they struggled to reconcile their religion with the obvious injustice of slavery. Some denied that Africans had souls, and others justified slavery with selected passages in the Bible. When some slaves adopted Christianity, Virginia passed a law declaring that their baptism "doth not exempt them from bondage."

However, Cotton Mather, a 17th-century New England minister, told his congregation, "Thy Negro is thy neighbor," and formed a society to educate Africans in Christianity. Pennsylvania Quakers, having come to America to escape religious persecution, were particularly strong in their condemnation of slavery and the slave trade. Free blacks, particularly in New England, also spoke out on behalf of their brothers and sisters in bondage.

The comparatively fine clothing of two members of this slave family indicates that they may have served in the "big house," the master's home.

Farm labor was not the sole occupation of African Americans in colonial America. Many, both slave and free, were blacksmiths, tailors, bakers, masons, carpenters, and other kinds of skilled workers. Africans were found among American ships' crews and dockworkers.

By the time the American colonies revolted against British rule in 1776, the population included nearly 700,000 people of African descent (about 20 percent of the total number of settlers), and some played a role in the Revolution. Probably the first American to give his life fighting British rule was Crispus Attucks, a former slave, who was killed in the Boston Massacre of 1770.

Though the British promised freedom to slaves who volunteered to fight the rebellious colonists, only about 1,000 accepted the offer. By contrast, at least 4,000 African Americans served in the army and navy on the patriot side. Two of them, Peter Salem and Salem Poor, distinguished themselves for bravery at the Battle of Bunker Hill.

The high ideals expressed in the Declaration of Independence ("All men are created equal...with certain unalienable rights, that among these are life, liberty, and the pursuit of happiness") clearly conflicted with the institution of slavery. Rhode Island, while still a colony, had banned the slave trade in 1774, and after independence more of the northern states followed suit. The Northwest Ordinance of 1787 prohibited slavery within the new western territories north of the Ohio River. However, the southern states, whose economy depended heavily on slave labor, resisted calls to abolish slavery. The anguished and bitter division between North and South had begun, and would remain a central theme of American history until the outbreak of the Civil War—and long afterward.

Slavery caused heated debates among the delegates who gathered to write the U.S. Constitution in 1787. They reached a compromise on the issue of counting the population to determine the number of representatives a state could send to Congress: each slave was counted as three-fifths of a free person. However, in another part of the Constitution, the antislavery forces won a victory—after 20 years, Congress could prohibit the importation of new slaves. Indeed, such a law was passed, and after 1807 no new slaves were brought into the United States legally. (The total number of Africans who arrived before that time was about 450,000.)

Some Americans believed that slavery would gradually cease without a formal ban, as it had in Massachusetts. The slave trade had declined ever since the 1760s, partly because the tobacco crop became less important in the original southern colonies. In Virginia alone, about 10,000 slaves were voluntarily freed by their masters in 1789–90.

However, the 1793 invention of the cotton gin, which processed raw cotton mechanically, created a new cash crop for the southern plantations. "King Cotton" became the South's major source of income, and because it required large numbers of cropworkers, slavery became ever more important and lucrative. Slavery spread as new southern states were admitted to the Union—Kentucky, Tennessee, Mississippi, Alabama, Missouri, Arkansas, Florida, and Texas. Louisiana, which was admitted in 1812, had practiced slavery since its foundation as a French colony.

What was slavery like? Only the slaves could answer this question, and fortunately, a great many of them provided their firsthand recollections, from about 1760 until the mid-20th century. Conditions varied greatly from place to place, depending on the humanity of the masters. In general, slaves were treated relatively better in the eastern and northern parts of the South than they were in the "Deep South." But everywhere in the slave states, the whim of a master was enough to forever separate husbands from wives at a slave auction. Parents could only watch helplessly as their children were sold to a new master. Slave owners who had raped their female slaves kept their own children in bondage. There was no possible appeal for the slaves, for southern state laws denied them any rights.

Slaveholders continually argued that most slaves were content with their condition. However, the harsh slave codes passed by all southern states belied this argument. These laws had one simple purpose: to keep slaves from escaping. For a slave, there was no such thing as freedom of speech, of assembly, or even of worship—the approval and sometimes the presence of a white was required for any of these activities. Slaves had to obtain passes before they could leave the plantation for any reason; organized groups of white "paddyrollers" (a corruption of the word *patrollers*) checked passes on the roads.

A group of slaves on a plantation on Edisto Island, South Carolina, in 1862. A year before, the war that would set them free had begun only a few miles north.

It was against the law for anyone to teach a slave to read or write—perhaps the clearest indication of the slave owners' fear that ideas of liberty might spread among the slaves. Yet despite every effort of southern society, there were a few large-scale slave rebellions, even in colonial times. The last of these, led by Nat Turner, took place in Virginia in 1831.

By that time, slaves seeking freedom had another way open to them—the Underground Railroad. Opponents of slavery, both whites and free blacks, began to organize a network of "stations" where escaped slaves could find refuge. "Conductors," many of whom were former slaves, risked their lives and liberty to return South and lead their brothers and sisters to freedom. On foot or in wagons, on ships and in railway cars, the passengers on the Underground Railroad found safety in little towns and large cities where people still believed that all were created equal.

A number of escaped slaves, such as Frederick Douglass, wrote autobiographies that became best-sellers. Henry Bibb, another passenger on the Railroad, started a newspaper called *The Voice of the Fugitive.* Many escapees gave lectures to northern audiences, giving the slave's version of slavery—and sometimes pointing out the racial discrimination that existed in the North as well. These men and women eloquently disproved the slaveholders' contention that Africans could not be educated. People of conscience gradually came to the conclusion that ending the evil of slavery was a cause that could no longer be denied.

In 1863, in the midst of the Civil War, President Lincoln issued the Emancipation Proclamation, which ended slavery in the states that had rebelled. Throughout the South, even before Union troops arrived, the word spread through the slave quarters. Felix Haywood, a slave, recalled that his people walked "on golden clouds."

On December 18, 1865, the 13th Amendment to the Constitution was ratified, ending slavery throughout the United States.

EXPLORERS AND PIONEERS

The earliest African American explorer of the United States was Estevanico, sometimes called Esteban or Stephan Dorantez. Estevanico came to Florida in 1528 with a 300-member Spanish expedition that traveled overland to present-day Texas. He was one of only four people who survived the harrowing six-year journey. Because Estevanico had been able to befriend the Native Americans, he was chosen to accompany Brother Marco de Nizza on a second exploring trip in 1539, into what is today Arizona. Brother Marco, a member of the Franciscan order, was searching for the fabled Seven Cities of Cíbola, said to be filled with gold and silver. Estevanico, whom Brother Marco calls Stephan, went ahead with a group of Native American guides, leaving Brother Marco behind. One of the guides soon returned with the news that Estevanico had discovered the Seven Cities and that Brother Marco should follow. Other messengers came back to report what happened to Estevanico. The following, in Brother Marco's words, describes Estevanico's fate.

Estevanico, the African who explored much of the southern part of the United States, from Florida to Arizona.

Here met us an Indian...which had gone before with Stephan, who came in a great fright, having his face and body all covered with sweat...and he told me that...before Stephan came to Cíbola he sent his great Mace made of a gourd, [which] had a string of bells upon it, and two feathers one white and other red, in token that he demanded safe conduct, and that he came peaceably. [But the man who received the gourd] took the same in his hands...in a great rage and fury...cast it to the ground, and [told] the messengers to get them packing with speed...in no case to enter into the city, for if they did he would put them all to death. [However, Estevanico ignored the warning and pressed onward] where he found men that would not let him enter into the town, but shut him into a great house...and they kept him there all that night without giving him meat or drink.

[Brother Marco hurried on toward Cíbola and in a few days] met two other Indians of those which went with Stephan, which were bloody and wounded...and...began to make great lamentation. [They said that] the next day when the sun was a lance high, Stephan went out of the house...and suddenly came [a great number] of people from the city...and forthwith they shot at us and wounded us...and after this we could not see Stephan any more, and we think they have shot him to death, as they have done all the rest which went with him, so that none are escaped but we only.

The names of African slaves are seldom given in the old Spanish chronicles of exploration. However, the Spaniards nearly always brought Africans with them—for they were needed to do the hard work. In 1565, Pedro Menéndez de Áviles led an expedition to Florida, where he founded a settlement called St. Augustine. Menéndez's instructions from the King of Spain, as recorded by a Spanish historian of the time, show the African role in what is the oldest city in today's United States.

He was to settle five hundred colonists—one hundred of them married couples and the majority of the others farmers and craftsmen—so that the earth might be the more easily cultivated. He was to take twelve religious and four priests of the Society of Jesus. During the same period he was to bring into Florida a hundred horses and mares, two hundred heifer-calves, four hundred hogs, four hundred sheep, a few goats, and all the livestock, both major and minor, that he wanted. Also, he would take five hundred slaves (for whom he would be given a duty-free license), a third of them females. They would be employed by himself and the people he took with him to build, populate, and cultivate Florida the more easily and to plant cane and construct sugar mills.

In the 1770s, Jean Baptiste Point du Sable, an African American, built the first permanent settlement on the site of today's city of Chicago. Mrs. John Kinzie, whose father-in-law purchased du Sable's trading post, provided one of the first accounts of this early settler.

In giving the early history of Chicago, the Indians say, with great simplicity, "the first white man who settled here was a negro."

This was Jean Baptiste Point-au-Sable, a native of St. Domingo, who...found his way to this remote region, and commenced a life among the Indians. There is usually a strong affection between these two races, and Jean Baptiste imposed upon his new friends by making them believe that he had been a "great chief" among the whites.

Because du Sable's name was French, he is believed to have been born in the French colony of Haiti. This French connection got him into trouble during the American Revolution, when France was an ally of the rebellious colonists. In 1779, a British military commander in the area ordered du Sable taken into custody.

I had the negro, Baptiste Point de Sable brought prisoner from the River Du Chemin. Corporal Tascon, who commanded the party, very prudently prevented the Indians from burning his home, or doing him any injury.... The negro, since his imprisonment has in every way behaved in a manner becoming to a man of his station, and has many friends who give him good character.

Jean Baptiste Point du Sable claimed U.S. citizenship after the Revolution. In 1783, he received a land grant in the Northwest Territory and built up his trading post, which eventually included a large house, a mill, a bakery, a dairy, a smokehouse, a workshop, a stable, and a barn.

SLAVE LIFE

In 1712, the colony of Carolina issued the first comprehensive slave code in North America. It begins:

Whereas, the plantations and estates of this province cannot be well and sufficiently managed and brought into use, without the labor and service of negroes and other slaves; and forasmuch as the said negroes and other slaves brought into the people of this Province for that purpose, are of barbarous, wild, savage natures, and such as renders them wholly unqualified to be governed by the laws, customs, and practices of this Province; but that it is absolutely necessary, that such other constitutions, laws and orders, should in this Province be made and enacted, for the good regulating and ordering of them, as may restrain the disorders, rapines and inhumanity, to which they are naturally prone and inclined, and may also tend to the safety and security of the people of this Province and their estates; to which purpose,

BE IT THEREFORE ENACTED...that all negroes, mulattoes, mestizoes or Indians, which at any time heretofore have been sold, or now are held or taken to be, or hereafter shall be bought and sold for slaves, are hereby declared slaves; and they, and their children, are hereby made and declared slaves, to all intents and purposes.

Around 1840, Solomon Northrup, a free African American, was kidnapped in Washington, D.C., and sold into slavery. In his book Twelve Years a Slave, *published in 1853, Northrup describes his life in the cotton fields of Louisiana.*

In the latter part of August begins the cotton picking season. At this time each slave is presented with a sack. A strap is fastened to it, which goes over the neck, holding the mouth of the sack breast high, while the bottom reaches nearly to the ground. Each one is also presented with a large basket that will hold about two barrels. This is to put the cotton in when the sack is filled.

An ordinary day's work is two hundred pounds. A slave who is accustomed to picking is punished if he or she brings in a less quantity than that.....

The hands are required to be in the cotton field as soon as it is light in the morning, and, with the exception of ten or fifteen minutes, which is given them at noon to swallow their allowance of cold bacon, they are not permitted to be a moment idle until it is too dark to see and when the moon is full they often times labor till the middle of the night. They do not dare to stop...however late it be, until the order to halt is given by the driver.

[After] the day's work in the field, the baskets are...carried to the gin-house, where the cotton is weighed. No matter how fatigued and weary he may be—no matter how much he longs for sleep and rest—a slave never approaches the gin-house with his basket of cotton but with fear. If it falls short in weight—if he has not performed the full task appointed him, he knows that he must suffer. And if he has exceeded it by ten or twenty pounds, in all probability his master will measure the next day's task accordingly....

This done, the labor of the day is not yet ended, by any means. Each one must then attend to his respective chores. One feeds the mules, another the swine—another cuts the wood, and so forth; besides, the packing is all done by candle light. Finally, at a late hour, they reach the quarters, sleepy and overcome with the long day's toil. Then a fire must be kindled in the cabin, the corn ground in a small hand-mill, and supper, and dinner for the next day in the field, prepared....

An hour before daylight the horn is blown. Then the slaves arouse, prepare their breakfast, fill a gourd with water, in another deposit their dinner of cold bacon and corn cake, and hurry to the field again. It is an offence invariably followed by a flogging, to be found at the quarters after daybreak. Then the fears and labors of another day begin; and until its close there is no such thing as rest.

What slavery was really like can be seen in the scars on the back of this former slave, whipped many times for violating the rules of the plantation.

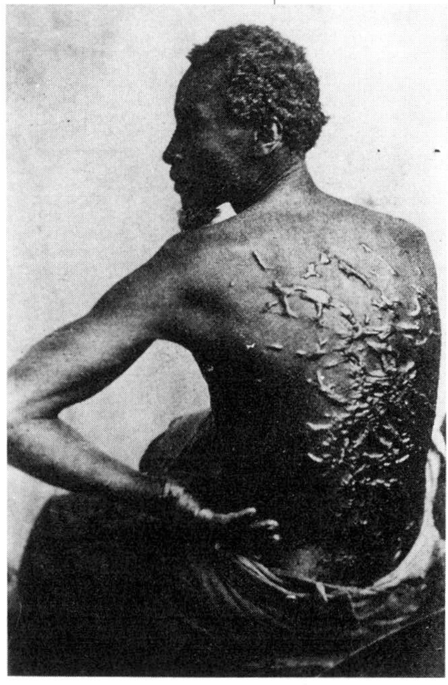

During the 1930s, members of the Federal Writers' Project traveled through the southern states and compiled life histories of former slaves. The spelling of their dialect was invented by the interviewers in an attempt to reproduce the speech of the ex-slaves. Katie Darling of Marshall, Texas, was 88 years old when she was interviewed.

You is talkin' now to a nigger what nursed seven white chillen in them bullwhip days. Miss Stella, my young missy, got all our ages down in the Bible, and it say I'se born in 1849.

Massa Bill McCarty my massa and he live east and south of Marshall, close to the Louisiana line.... Massa have six chillen when war come on and I nursed all of 'em. I stays in the house with 'em and slept on a pallet on the floor, and soon I'se big 'nough to tote the milk pail and they puts me to milkin', too. Massa have more'n one hundred cows and most of the time me and Violet do all the milkin'. We better be in that cow pen by five o'clock. One mornin' Massa cotched me lettin' one the calves do some milkin' and he let me off without whippin' that time, but that don't mean he always good, 'cause them cows have more feelin' for us than Massa and Missy.

We et peas and greens and collards and middlin's. Niggers had better let that ham alone! We have meal coffee. They parch meal in the oven and boil it and drink the liquor....

At night the men had to shuck corn and the women card and spin [cloth]. Us got two pieces of clothes for winter and two for summer, but us have no shoes. We had to work Saturday all day and if that grass was in the field we didn't get no Sunday, either.

They have dances and parties for the white folks' chillen, but Missy say, "Niggers was made to work for white folks,"

The most important crop of the slave states was cotton. During the harvest, the field slaves carried their day's picking to be weighed. If they failed to meet their quota, they were punished.

An unknown artist painted this plantation scene sometime between 1790 and 1800. Scholars originally thought it to be a depiction of a wedding ceremony in which slaves jumped over a broomstick, but some now believe it perhaps depicts a secular West African dance performed with scarves and sticks.

At the outbreak of the Civil War in 1861, about 5 percent of southern whites owned slaves. But most nonslaveholders shared the prejudice against African Americans. A northern visitor, Frederick Law Olmsted, asked a white Alabama farmer before the Civil War what he thought about freeing the slaves:

Well, I'll tell you what I think on it; I'd like it if we could get rid of 'em.... I wouldn't like to hev 'em freed, if they was gwine to hang 'round. They ought to get some country and put 'em war they could be by themselves. It wouldn't do no good to free 'em, and let 'em hang 'round, because they is so monstrous lazy; if they hadn't got nobody to take keer on 'em, you see they wouldn't do nothin' but juss nat'rally laze 'round, and steal, and pilfer, and no man couldn't live, you see, war they was—if they was free, no man couldn't live—and this ere's the other. Now suppose they was free, you see they'd all think themselves just as good as we, of course they would, if they was free. Now, just suppose you had a family of children, how would you like to hev a niggar steppin' up to you darter? Of course you wouldn't, and that's the reason I wouldn't like to hev 'em free; but I tell you, I don't think it's right to hev 'em slaves so; that's the fac—taant right to keep 'em as they is.

Slave quarters on a plantation on Fort George Island, Florida. The photo may have been taken on Sunday, the one day of the week when slaves were usually allowed to rest.

and on Christmas Miss Irene bakes two cakes for the nigger families but she doesn't let Missy know about it.

When a slave die, Massa make the coffin hisself and send a couple niggers to bury the body and say, "Don't be long," and no singin' or prayin' allowed, just put them in the ground and cover 'em up and hurry on back to that field.

At the age of 93, Sam Polite recalled his slave life on one of the Sea Islands off the coast of South Carolina.

When gun shoot on Bay Point for Freedom [in 1865], I been seventeen-year-old working slave. I born on B. Fripp Plantation, on St. Helena Island. My father belong to Mr. Old B. Fripp. I don't know how mucher land...he have, but he have...more'n a hundred slave.

Slave live on street—two rows of house with two room to the house....

When I been little boy, I play on street—shoot marble, play army, and such thing. When horn blow and morning star rise, slave have for get up and cook. When day clean [after sunrise], they gone to field. Woman too old for work in field have for stay on street and mind baby. Old mens follow cow. Chillun don't work in field till twelve or thirteen year old. You carry dinner to field in your can and leave 'um at the heading [top of row]. When you feel hungry, you eat....

When you knock off work, you can work on your land. Maybe you might make two or three task of land round your cabin what Marster give you for plant. You can have chicken, maybe hog. You can sell egg and chicken to store and Marster will buy your hog. In that way, slave can have money for buy thing like fish and whatever he want. We don't get much fish in slavery, 'cause we never have boat. But sometime you can throw out net and catch shrimp. You can also catch possum and raccoon with your dog.

On Saturday night, every slave that works gets peck of corn and pea, and sometime meat and clabber. You never see any sugar, neither coffee, in slavery. You has straw in your mattress, but they gibe you blanket. Every year, in Christmas month, you get four or either five yard cloth, according to how [big] you is. Out of that, you have to make your clote [clothes]. You wears that same clote till the next year. You wear it winter and summer, Sunday and every day. You don't get no coat, but they give you shoe.

In slavery, you don't know nothing about sheets for your bed. Us never know nothing about Santa Claus till Freedom, but on Christmas, Marster give you meat and syrup and maybe three day without work.

Slave work till dark on Saturday just like any other day. I still does work till dark on Saturday. But on Sunday, slave don't work. On Fourth of July, slave work till twelve o'clock and then knocks off. On Sunday, slave can visit back and forth on the plantations.

Slaves who did field work envied those who worked in "the big house" as personal servants to the master's family. But big house slaves also had to work hard, as one woman told an interviewer in 1837.

When I was nine years old, they took me from my mother and sold me. Massa Tinsley made me the house girl. I had to make the beds, clean the house, and other things. After I finished my regular work, I would go to the mistress's room, bow to her, and stand there till she noticed me. Then she would say, "Martha, are you through with your work?" I'd say, "Yes mam." She'd say, "No you ain't; you haven't lowered the shades." I'd lower the shades, fill the water pitcher, arrange the towels on the washstand, and anything else mistress wanted me to do. Then she'd tell me that was about enough to do in there. Then I would go to the other rooms in the house and do the same things. We weren't allowed to sit down. We had to be doing something all day. Whenever we were in the presence of any of the white folks, we had to stand up.

Physical punishment was always a part of slavery. Elizabeth Keckley, born in Virginia in the 1820s, described being whipped by an overseer.

He seized a rope, caught me roughly, and tried to tie me. I resisted with all my strength, but he was the stronger of the two, and after a hard struggle succeeded in binding my hands and tearing my dress from my back. Then he picked up a rawhide, and began to ply it freely over my shoulders. With a steady hand and practised eye he would raise the instrument of torture, nerve himself for the blow, and with fearful force the rawhide descended upon the quivering flesh. It cut the skin, raised great welts, and the warm blood trickled down my back. Oh God! I can feel the torture now—the terrible, excruciating agony of those moments. I did not scream; I was too proud to let my tormentor know what I was suffering. I closed my lips firmly, that not even a groan might escape from them, and I stood like a statue while the keen lash cut deep into my flesh.

Frederick Douglass, born a slave in Maryland around 1818, escaped when he was 19 and became an eloquent spokesman for the antislavery movement. In his autobiography, Douglass wrote of the cruelty of the slave masters and mistresses.

I speak advisedly when I say this,—that killing a slave, or any colored person, in Talbot county, Maryland, is not treated as a crime, either by the courts or the community. Mr. Thomas Lanman...killed two slaves, one of whom he killed with a hatchet, by knocking his brains out. He used to boast of the commission of the awful and bloody deed.... The wife of Mr. Giles Hick, living but a short distance from where I used to live, murdered my wife's cousin, a young girl between 15 and 16 years of age.... The offence for which this girl was thus

As Others Saw Them

A German immigrant, Gustav Koerner, traveled through Missouri and noted the condition of slavery:

Passing the court house, we saw colored men, women and children sold at auction. We were also shown a sort of prison, where refractory slaves were confined at the request of their masters or were whipped at their masters' cost, by men regularly appointed for that purpose.... From the second story of our residence we could see into the yard of a neighboring house, where we once saw what appeared to be an American lady, lashing a young slave girl with a cowhide. Had there still been a lingering disposition in the Engelmann or Abend family to settle in Missouri, these scenes would have quenched it forever.

A collection of metal tags that African Americans in the slave states were required to wear. Some indicated their occupations; the one at top right showed that the bearer was free.

murdered was this:—She had been set that night to mind Mrs. Hick's baby, and during the night she fell asleep, and the baby cried. She, having lost her rest for several nights previous, did not hear the crying. They were both in the room with Mrs. Hicks. Mrs. Hicks, finding the girl slow to move, jumped from her bed, seized an oak stick of wood by the fireplace, and with it broke the girl's nose and breastbone, and thus ended her life.

During the California gold rush, southern whites took their slaves to the goldfields. When California became a state in 1850, slavery was banned, and some slaves claimed or purchased their freedom. The state constitution, however, forbade "Indians, Africans, and the descendants of Africans" from voting.

Former slave Austin Steward wrote his autobiography in 1857. Though he was then 62, he recalled vividly the last time he had seen his father, over half a century earlier, on "Mr. N's" farm.

The only incident I can remember, which occurred while my mother continued on N's farm, was the appearance of my father one day with his head bloody and his back lacerated. He was in a state of great excitement, and though it was all a mystery to me at the age of three or four years, it was explained at a later period, and I understood that he had been suffering the cruel penalty of the Maryland law for beating a white man. His right ear had been cut off close to his head, and he had received a hundred lashes on his back. He had beaten the overseer for a brutal assault on my mother, and this was his punishment. Furious at such treatment, my father became a different man, and was so morose, disobedient, and intractable, that Mr. N. determined to sell him. He accordingly parted with him, not long after...and neither mother nor I ever heard of him again.

After the slave trade was banned in 1807, slave owners derived profits from breeding their slaves and selling their children. In the 1970s, Mary Murphy, born in 1894 in Missouri, told an interviewer about her grandmother's experiences.

My grandmother on my father's side was a black African. She was born and reared in Louisville, Kentucky, on a big plantation. Her boss's name was Larsh. My grandmother was [one] of eight children. When she got to be sixteen years old my great-grandmother put a long dress on her. That let the white man know that she's coming to womanhood and she's eligible to be placed in a house, or a cabin, as they called it, to raise children. Mr. Larsh came by one Saturday and said to great-grandmother, "Hannah, I see you got a long dress on Marianne."

She said, "Yes, she's comin' to womanhood now."

He said, "Well, I've got a youngster over here that's a nice strapping young man. I want to put them in a cabin and raise me some young'uns."

Now, my great-grandmother hated to give her up but she did. My grandmother had eight children by this man.

And then this same boss man came by again. He said, "Marianne, give your baby to Sara"—that was her other sister—"and come and go with me." She gave her baby to Sara and came to the door. Then they called jackets a bass. He said, "You'd better get a bass and put [it] on." So she reached back

On a Georgia plantation, a girl helps sharpen an axe. As soon as slave children were able, they were required to work. If their master chose to, he could sell them away from their parents.

and got her a little jacket to put on. When she got outside there was ninety-nine other people outside. They were going on this travel from Louisville, Kentucky, to St. Louis, Missouri. They didn't know it, but they were going there to be sold.... My grandmother...walked all the way. They walked in the daytime and then they'd rest at night.... When they got to St. Louis, my grandmother's breasts had swollen and were so sore they put her on the block first. She tried to step up on the block and she couldn't. So they helped her up on the block because her bust was swollen so. The boss man tore her clothes all off to show how big she was. She said she just cried. They sold her for one thousand dollars. They sold her to this man, George Lucas.

He and his wife took her immediately to their plantation out in Missouri because she was suffering. They put her with another Negro man. His name—everybody that worked for this man, their last name was Lucas—was Steven Lucas. That was my father's father. They had eight children, so that made my grandmother the mother of sixteen children. We didn't know her real age, but as well as we could count she lived to be ninety-five years old before she passed. She was living in Oklahoma when she passed. Slavery was over.

Sometimes the slaves could choose a husband or wife. They would conduct a marriage ceremony, usually by jumping over a broomstick. Sometimes, a black minister officiated at the wedding, but more often, there was no minister. During the 1930s, an old woman who had been a slave described the experience.

Didn't have to ask Marsa or nothin'. Just go to Aunt Sue an' tell her you want to git mated. She tell us to think 'bout it hard fo' two days, 'cause marryin' was sacred in de eyes of Jesus. Arter two days Mose an' I went back an' say we done thought 'bout it an' still want to git married. Den she called all de slaves arter tasks to pray fo' de union dat God was gonna make. Pray we stay together an' have lots of chillun an' none of 'em git so' way from de parents. Den she lay a broomstick 'cross de sill of de house we gonna live in an' jine our hands together. Fo' we step over it she ast us once mo' if we sho' we wanted to git married. 'Course we said yes. Den she say, "In de eyes of Jesus step into Holy land of mat-de-money." When we step 'cross the broomstick, we was married. Was bad luck to tech de broomstick. Fo'ks always stepped high 'cause they didn't want no spell cast on 'em—Aunt Sue used to say whichever one teched de stick was gonna die fust.

Treated as property, slaves could at any time be separated from their husbands, wives, parents, or children. "Old Elizabeth," a former slave, told her life story in 1873, when she was 97.

I was born in Maryland in the year 1766. My parents were slaves. Both my father and mother were religious people, and belonged to the Methodist Society. It was my father's practice to read in the Bible aloud to his children every sabbath morning....

Usually, slaves received new clothing and shoes only once a year, which explains the patched and ragged clothes of this group. Most masters provided only enough food and shelter to keep their slaves healthy enough to work.

Many state legislatures in the South passed laws preventing anyone from teaching slaves to read or write. A member of the Virginia legislature explained the reason for such laws:

We have, as far as possible, closed every avenue by which light might enter their minds. If you could extinguish the capacity to see the light, our work would be completed; they would then be on a level with the beasts of the field, and we should be safe!

In the eleventh year of my age, my master sent me to another farm, several miles from my parents, brothers, and sisters, which was a great trouble to me. At last I grew so lonely and sad I thought I should die, if I did not see my mother. I asked the overseer if I might go, but being positively denied, I concluded to go without his knowledge. When I reached home my mother was away. I set off and walked twenty miles before I found her. I staid with her for several days, and we returned together. Next day I was sent back to my new place, which renewed my sorrow. At parting, my mother told me that I had "nobody in the wide world to look to but God." These words fell upon my heart with pondrous weight, and seemed to add to my grief. I went back repeating as I went, "none but God in the wide world." On reaching the farm, I found the overseer was displeased at me for going without his liberty. He tied me with a rope, and gave me some stripes of which I carried the marks for weeks.

Moses Grandy, a former slave, described his mother's efforts to keep her children from being sold to another plantation.

The master, Billy Grandy, whose slave I was born, was a hard drinking man; he sold away many slaves. I remember four sisters and four brothers; my mother had more children, but they were dead or sold away before I can remember. I was the youngest. I remember well my mother often hid us all in the woods, to prevent master selling us. When we wanted water, she sought for it in any hold or puddle, formed by falling trees or otherwise: it was often full of tadpoles and insects: she strained it, and gave it round to each of us in the hollow of her hand. For food, she gathered berries in the woods, got potatoes, raw corn, &c. After a time the master would send word to her to come in, promising he would not sell us. But, at length, persons came, who agreed to give the prices he set on us. His wife...prevailed on him not to sell me; but he sold my brother, who was a little boy. My mother, frantic with grief, resisted their taking her child away; she was beaten and held down: she fainted, and when she came to herself, her boy was gone. She made much outcry, for which the master tied her up to a peach tree in the yard, and flogged her.

During the Civil War, the owner of these South Carolina slaves fled his plantation as the Union army approached. Freed slaves, known as "contraband," joined the Union forces, served as cooks and hospital workers, or took shelter in camps that the army established.

Many slave owners raped their female slaves. Jack Maddox described what happened when his master, Judge Maddox, brought home a young female slave.

Judge Maddox was buying a slave every now and then. One day he brought home a pretty mulatto gal. She was real bright and she had long black straight hair and was dressed neat and good. The old lady [the Judge's wife] came out of the house and took a look and said, "What you bring that thing here for?" The judge said, "Honey, I brung her here for you. She going to do your fine needlework." She said, "Fine needlework, your hind leg!"

Well, you know what that old lady done? When Judge Maddox was away from home she got the scissors and cropped that gal's head to the skull. I didn't know no more 'bout that case, but one thing I do know was that white men got plenty chilluns by the nigger women. They didn't ask them. They just took them. I heard plenty 'bout that.

Slave owners tried to keep their slaves from learning how to read or write. James Williams, who had been a slave on the estate of George Larrimore, recalled what happened when he showed an aptitude for learning.

Mr. Larrimore had three children; George, Jane, and Elizabeth. The former was just ten days older than myself; and I was his playmate and constant associate in childhood. I used to go with him to his school, and carry his books for him as far as the door, and meet him there when the school was dismissed. We were very fond of each other, and...he taught me the letters of the alphabet, and I should soon have acquired a knowledge of reading, had not George's mother discovered her son in the act of teaching me. She took him aside and severely reprimanded him. When I asked him, not long after, to tell me more of what he had learned at school, he said that his mother had forbidden him to do so any more, as her father had a slave who was instructed in reading and writing, and on that account proved very troublesome. He could imitate the hand-writing of all the neighboring planters, and used to write passes and certificates of freedom for the slaves, and finally wrote one for himself, and went off to Philadelphia, from whence her father received from him a saucy letter, thanking him for his education.

Though the slave states had laws that forbade anyone to teach the slaves to read and write, many took the risk. Susie King Taylor, who later served as a nurse for the first African American regiment in the Civil War, described how she learned to read and write.

I was born under the slave law in Georgia, in 1848, and was brought up by my grandmother in Savannah.... My brother and I being the two eldest, we were sent to a friend of my grandmother, Mrs. Woodhouse, a widow, to learn to read and write. She was a free [black] woman and lived...about half a mile from my house. We went every day about nine o'clock, with our books wrapped in paper to prevent the police or white persons from seeing them. We went in, one at a time, through the gate, into the yard to the kitchen, which was the schoolroom. She had 25 or 30 children whom she taught, assisted by her daughter, Mary Jane. The neighbors would see us going in sometimes, but they supposed we were there learning trades, as it was the custom to give children a trade of some kind. After school we left the same way we entered, one by one, when we would go to a square, about a block from the school and wait for each other.

Frederick Douglass recalled a song that gave what he called "not a bad summary of the palpable injustice and fraud of slavery, giving, as it does, to the lazy and the idle the honest comforts which God designed should be given solely to the honest laborer":

We raise the wheat,
They give us the corn;
We bake the bread.
They give us the crust;
We sift the meal,
They give us the husk;
We peel the meat,
They give us the skin;
And that's the way,
They take us in;
We skim the pot,
They give us the liquor;
And say that's good enough for nigger.
Your butter and the fat;
Poor nigger, you can't ever get that!
Walk over—

Sojourner Truth

Some of the women who supported the abolitionist cause were also active in the women's rights movement of the time. At a women's rights convention in May 1851, the delegates were surprised when a black woman approached the platform. She was Sojourner Truth, a former slave in New York State. Having just heard a male minister at the convention declare that women were inferior to men, Truth felt she had to speak out. As another delegate to the convention recalled:

"At her first word there was a profound hush. She spoke in deep tones, which, though not loud, reached every ear in the house....

'That man over there say that women need to be helped into carriages, and lifted over ditches, and to have the best place everywhere. Nobody ever helps me into carriages, or over mud-puddles, or gives me any best place!' And, raising herself to her full height, and her voice to a pitch like rolling thunder, she asked: 'And ain't I a woman? Look at me! Look at my arm!' (and she bared her right arm to the shoulder, showing her tremendous muscular power). 'I have plowed, and planted, and gathered into barns, and no man could head me! And ain't I a woman? I could work as much and eat as much as a man—when I could get it—and bear the lash as well! And ain't I a woman? I have borne thirteen children, and seen 'em most all sold off to slavery, and when I cried out with my mother's grief, none but Jesus heard me! And ain't I a woman?'"

FREE DURING SLAVERY

A small minority of African Americans were legally free during the colonial period and in the United States before Emancipation. One of them was Benjamin Banneker, a scientist who was one of the surveyors of the land on which Washington, D.C., was built. Banneker wrote to Thomas Jefferson to rebuke the author of the Declaration of Independence for declaring "all men are created equal" while he himself owned slaves.

[August 19, 1791]

Sir, suffer me to recall to your mind that time, in which the arms of the British crown were exerted, with every powerful effort, in order to reduce you to a state of servitude: look back, I entreat you...you were then impressed with proper ideas of the great violations of liberty, and the free possession of those blessings, to which you were entitled by nature; but, Sir, how pitiable is it to reflect that although you were so fully convinced of the benevolence of the Father of Mankind, and of his equal and impartial distribution of these rights and privileges which he hath conferred upon them, that you should at the same time counteract his mercies, in detaining by fraud and violence, so numerous a part of my brethren under groaning captivity and cruel oppression, that you should at the same time be found guilty of that most criminal act, which you professedly detested in others.

Nancy Prince was born free in 1799 in Newburyport, Massachusetts. Her family was poor and Nancy was in domestic service from the age of eight. Later in life, she met and married Prince, a sailor who had served in the Russian czar's court. They sailed to Russia, where Nancy Prince lived for nine years before returning home because she could not stand the cold weather. Years later she wrote a book about her life. She described the drudgery of her work in service when she was only 14 years old.

There were seven in the family, one sick with fever, and another in a consumption [tuberculosis]; and of course, the work must have been very severe, especially the washings. Sabbath evening I had to prepare for the wash; soap the clothes and put them into the steamer, set the kettle of water to boiling, and then close in the steam, and let the pipe from the boiler into the steam box that held the clothes. At two o'clock, on the morning of Monday, the bell was rung for me to get up; but, that was not all, they said I was too slow, and the washing was not done well; I had to leave the tub to tend the door and wait on the family, and was not spoken kind to, at that.

Hard labor and unkindness was too much for me; in three months, my health and strength were gone. I often looked at my employers, and thought to myself, is this your religion? I did not wonder that the girl who had lived there previous to myself went home to die. They had family prayers, morning and evening. Oh! yes, they were sanctimonious! I was a poor stranger, but fourteen years of age, imposed upon by these good people; but I must leave them. In the year 1814, they sent me to Gloucester in their chaise. I found my poor mother in bad health, and I was sick also; but, by the mercy of God, and the attention and skill of Dr Dale, and the kindness of friends, I was restored, so that in a few months, I was able again to go to work, although my side afflicted me, which I attributed to overworking myself.

The Anti-Slavery Almanac of 1839 printed this drawing of a free black man in the North captured to be sold into slavery.

David Walker was born free in North Carolina in the late 18th century. After years in the South, he went to Boston, where he opened a secondhand clothing store. Here, he learned to read and took up the cause of abolition of slavery. In addition, he criticized the attempts to send African Americans back to Africa, arguing that they were as fully American as the white population. In 1829, he issued Walker's Appeal.

Will any of us leave our homes and go to Africa? I hope not. Let them commence their attack upon us as they did on our brethren in Ohio, driving and beating us from our country, and my soul for theirs, they will have enough of it. Let no man of us budge one step, and let slaveholders come to beat us from our country. America is more our country, than it is the whites'—we have enriched it with our blood and tears. The greatest riches in all America have arisen from our blood and tears:—and will they drive us from our property and homes, which we have earned with our blood?...

Throw away your fears and prejudices then, and enlighten us and treat us like men, and we will like you more than we do now hate you; and tell us now no more about colonization, for America is as much our country, as it is yours.—Treat us like men, and there is no danger but we will all live in peace and happiness together. For we are not like you, hardhearted, unmerciful, and unforgiving. What a happy country this will be, if the whites will listen.

This painting shows the death of Crispus Attucks, the free African American who was killed by British troops in the Boston Massacre of 1770.

Free African Americans both in the North and the South often were deprived of such basic rights as voting and owning property. Everywhere they faced discrimination. Washington Spaulding, a free black Kentuckian, described the perils of being African American in the first half of the 19th century.

Our principal difficulty here grows out of the police laws, which are very stringent. For instance, a police officer may go [to] a house at night, without any search warrant, and, if the door is not opened when he knocks, force it in, and ransack the house, and the coloured man has no

As Others Saw Them

The Frenchman Alexis de Tocqueville visited the United States in the 1830s. He described the separation of blacks and whites in the free North:

In the theaters gold cannot procure a seat for the servile race beside their former masters; in the hospitals they lie apart; and although they are allowed to invoke the same God as the whites, it must be at a different altar and in their own churches, with their own clergy. The gates of heaven are not closed against them, but their inferiority is continued to the confines of the other world.

Black and white Americans formed antislavery organizations that helped slaves escape to the North. Some groups raised money to purchase slaves' freedom. Copies of this photograph of emancipated slaves from Louisiana were sold to raise funds to open schools for blacks.

redress. At other times, they come and say they are hunting for stolen goods or runaway slaves, and, some of them being great scoundrels, if they see a piece of goods, which may have been purchased, they will take it and carry it off. If I go out of the state, I cannot come back to it again. Then penalty is imprisonment in the penitentiary.... If a freeman comes here (perhaps he may have been born free), he cannot get free papers, and if the police find out that he has got no free papers, they snap him up, and put him in jail. Sometimes they remain in jail three, four, and five months before they are brought to trial. My children are just tied down here. If they go to Louisiana, there is no chance for them, unless I can get some white man to go to New Orleans and swear they belong to him, and claim them as his slaves.... There are many cases of assault and battery in which we can have no redress. I have known a case here where a man bought himself three times. The last time, he was chained on board a boat, to be sent South, when a gentleman who now lives in New York saw him, and bought him, and gave him his free papers.

Charlotte Forten was born in Philadelphia in 1837, the daughter of free African American abolitionists. As a teenager she moved to Salem, Massachusetts, to continue her education. She studied to be a teacher at the Salem Normal School and went on to become the first African American teacher in the Salem school system. In her diary, she confided some of the difficulties that African Americans met in the North.

Wednesday, Sept. 12 [1855].

Today school commenced—Most happy am I to return to the companionship of my studies,—ever my most valued friends. It is pleasant to meet the scholars [her pupils] again; most of them greeted me cordially, and were it not for the thought that *will* intrude, of the want of *entire sympathy* even of those I know and like best, I should greatly enjoy their society. There is one girl and only one—Miss [Sarah] B[rown] who I believe thoroughly and heartily appreciates anti-slavery,—*racial* anti-slavery, and has no prejudice against color. I wonder that every colored person is not a misanthrope. Surely we have everything to make us hate mankind. I have met girls in the schoolroom [who] have been thoroughly kind and cordial to me,—perhaps the next day met them in the street—they feared to recognize me; these I can but regard now with scorn and contempt,—once I liked them, believing them incapable of such meanness. Others give the most distant recognition possible,—I, of course, acknowledge no such recognitions, and they soon cease entirely. These are but trifles, certainly, to the great, public wrongs which we as a people are obliged to endure. But to those who experience them, these apparent trifles are most wearing and discouraging; even to the child's mind they reveal volumes of deceit and heartlessness, and early teach a lesson of suspicion and distrust. Oh! it is hard to go through life meeting contempt with contempt, hatred with hatred, fearing, with too good reason, to love and trust hardly any one whose skin is white—however lovable, attractive and congenial in seeming. In the bitter, passionate feelings of my soul again and again there rises the questions "When oh! when shall this cease?" "Is there no help?" "How long oh! how long must we continue to suffer—to endure?" Conscience answers it is wrong, it is ignoble to despair; let us labor earnestly and faithfully to acquire knowledge, to break down the barriers of prejudice and oppression. Let us take courage; never ceasing to work—hoping and believing that if not for us, for another generation there is a better, brighter day in store—when slavery and prejudice shall vanish before the glorious light of Liberty and Truth; when the rights of every colored man shall everywhere be acknowledged and respected, and he shall be treated as a *man* and a *brother*!

Some slaves managed to free themselves through a good deed to the community. Such a story was part of the heritage of N. Rebecca Taylor of Plainfield, New Jersey, who spoke to an interviewer in 1986.

I am seventy-six years old. I was born in Clinton, South Carolina. That's in Laurens County....

On my father's side my great-grandparents came from Pickens County in South Carolina. They were slaves. My great-grandfather had smallpox when he was a boy, so when he was grown he was sent to Newberry, South Carolina, to help take care of people during a smallpox epidemic. After the epidemic was over, out of gratitude, his master gave him his freedom.

He bought the freedom of my great-grandmother, who lived on the adjoining plantation, and they got married. They moved to Newberry and then to Laurens. He was a barber; he barbered for whites just like my father did. So he and my great-grandmother stayed in Laurens and raised all of their children there.

In the 19th century, more than 6,000 ex-slaves wrote their life stories, or related them to others for publication. One was Noah Davis, who made the following declaration in the preface to his 1859 autobiography.

The object of the writer, in preparing this account of himself, is to RAISE SUFFICIENT MEANS TO FREE HIS LAST TWO CHILDREN FROM SLAVERY. Having already, within twelve years past, purchased himself, his wife, and five of his children, at a cost, altogether, of over FOUR THOUSAND DOLLARS, he now earnestly desires a humane and christian public to AID HIM IN THE SALE OF THIS BOOK, for the purpose of finishing the task in which he has so long and anxiously labored.

Soon after the Revolutionary War, a group of abolitionists established the New York African Free School. In 1828, the school's principal sent examples of the students' work to an abolitionist convention.

On Freedom

Freedom will break the tyrant's chains,
And shatter all his whole domain;
From slavery she will always free
And all her aim is liberty.

—Thomas S. Sidney, age 12

Eliver Reason's Essay

Gentlemen,

I now address you in behalf of myself and my schoolmates:...

How many years have our poor Africans been in chains of slavery and perhaps have not seen a day of rest in many years, how likely is it, that they have been stolen from their native country, when they were young, from their dear father and mother; there are so many in the southern States chained in slavery for no other crime, than the color of their skin!... May the Supreme Being reward you ten fold for the good you do for us, is the desire of an injured African.

—Eliver Reason

This engraving of the New-York African Free-School No. 2 was based on a drawing made by a 13-year-old pupil. In 1824, New York City began to provide financial support for such schools, making it the first U.S. city to provide free education for African Americans.

Harriet Tubman

Harriet Tubman was the best known "conductor" on the Underground Railroad, a loosely organized chain of houses, or "stations," where escaped slaves could find refuge on their way to the North.

Born a slave in Maryland about 1821, Tubman was struck in the head by a rock thrown by her master when she was 13. For the rest of her life she suffered from dizzy spells and blackouts caused by the injury.

When she was about 25, Tubman escaped to freedom. However, she risked capture many times by returning to the South to rescue others from bondage. She became a legendary figure, known as Moses on the plantations where slaves spread the word that she was coming.

After passage of the Fugitive Slave Act of 1850 made it illegal to aid runaway slaves, she extended her route to Canada, where slavery had been abolished. "I don't trust Uncle Sam with my people any more," she declared.

During the Civil War, Tubman worked as a nurse, a scout, and a spy for Union troops. Afterward, she worked for the women's rights movement and lived in Auburn, New York, until her death in 1913.

ESCAPES AND REBELLIONS

Slaves continually attempted to escape from their masters, and many succeeded. On a few occasions, slaves organized full-scale rebellions that were crushed with great ferocity. One of the largest of these rebellions was led by Nat Turner in Virginia in 1831. Born in 1800, Turner had a brilliant mind and great personal magnetism. His mother told him that he "was intended for some great purpose." As an adult Turner preached that he was destined to lead the slaves out of bondage. On August 21, Turner and four others killed his master and armed themselves. About 60 other slaves eventually joined Turner's band, killing 55 whites over the next two days. The state militia put down the rebellion, but Turner managed to elude capture for about six weeks. Sentenced to death, Turner described his experiences to Thomas R. Gray, a lawyer, while awaiting execution. Gray reported that "for natural intelligence...[Turner] is surpassed by few men I have ever seen.... [He had a] mind capable of attaining anything."

Turner described his early years and the spiritual revelation that later came to him.

The manner in which I learned to read and write, not only had great influence on my own mind, as I acquired it with the most perfect ease, so much so, that I have no recollection whatever of learning to the alphabet—but to the astonishment of the family [of his master], one day, when a book was shewn to me to keep me from crying, I began spelling the names of different objects...and this learning was constantly improved at all opportunities.... Whenever an opportunity occurred of looking at a book...I would find many things that the fertility of my own imagination had depicted to me before; all my time, not devoted to my master's service, was spent either in prayer, or in making experiments in casting different things in moulds made of earth in attempting to make paper, gun-powder, and many other experiments, that although I could not perfect, yet convinced me of its practicability if I had the means. [When questioned as to the manner of manufacturing these different articles, he was found well informed on the subject.—Gray]....

Having soon discovered to be great, I must appear so, and therefore studiously avoided mixing in society, and wrapped myself in mystery, devoting my time to fasting and prayer....As I was praying one day at my plough, the spirit spoke to me, saying "Seek ye the kingdom of Heaven and all things shall be added unto you."

Question: What do you mean by the Spirit.

Answer: The Spirit that spoke to the prophets in former days—and I was greatly astonished and for two years prayed continually...and then again I had the same revelation, which fully confirmed me in the impression that I was ordained for some great purpose in the hands of the Almighty....

Question: Do you not find yourself mistaken now?

Answer: Was not Christ crucified?

James Williams grew up on a Virginia plantation in the early 1800s. After being sent to Alabama in the 1830s, James Williams found that slaves there were treated much more cruelly than on his old plantation. The white overseer forced him to whip other slaves. Williams found this intolerable, and one day seized the opportunity to escape.

I knew that a hot pursuit would be made after me, and what I most dreaded was that the overseer would procure Crop's bloodhounds to follow my track. [Crop was a professional slave-catcher.] If only the hounds of our plantation were sent after me, I had hopes of being able to make friends of them, as they were always good-natured and obedient to me. I travelled until, as near as I could judge, about ten o'clock, when a distant sound startled me. I stopped and listened. It was the deep bay of the bloodhound, apparently at a great distance.... I thought of the fate of Little John, [a runaway slave] who had been torn in pieces by the hounds, and of the scarcely less dreadful condition of those who had escaped the dogs only to fall into the hands of the overseer. The yell of the dogs grew louder. Escape seemed impossible. I ran down to the creek with a determination to drown myself. I plunged into the water and went down to the bottom, but the dreadful strangling sensation compelled me to struggle up to the surface.... As I rose to the top of the water and caught a glimpse of the sunshine and the trees, the love of life revived in me. I swam to the other side of the creek, and forced my way through the reeds to a large basswood tree, and stood under one of its lowest limbs, ready...to spring up into it. Here, panting and exhausted, I stood waiting for the dogs. The woods seemed full of them. I heard a bell tinkle, and, a moment after, our old hound Venus came bounding through the cane, dripping wet from the creek.... I called to her as I used to do when out hunting with her. She stopped suddenly, looked up at me, and then came wagging her tail and fawning around me. A moment after the other dogs came up hot in the chase.... I was just about to spring into the tree to avoid them, when Venus...met them, and stopped them.... The very creatures whom a moment before I had feared would tear me limb from limb, were now leaping and licking my hands, and rolling on the leaves around me. I listened awhile in the fear of hearing the voices of men following the dogs, but there was no sound in the forest save the gurgling of the sluggish waters of the creek, and the chirp of black squirrels in the trees. I took courage and started onward once more.

A group of escaped slaves who settled in Windsor, Ontario, in Canada, across the border from Detroit.

In 1859, John Brown, a white abolitionist, led a raid on the federal armory in Harpers Ferry, Virginia. He hoped to start a slave rebellion and establish a sanctuary for escaped slaves in the Appalachian Mountains. Brown succeeded in taking the armory, but he and his small band of followers were soon captured or killed. Brown was hanged in December, but his exploit thrilled many who wished all African Americans to be free. During the Civil War, northern troops marched into battle singing the song "John Brown's Body."

While Brown was awaiting execution, he received letters from sympathizers throughout the country. One came from an African American in Indiana:

Kendalville, Indiana, Nov. 25.
Dear Friend:

Although the hands of Slavery throw a barrier between you and me, and it may not be my privilege to see you in your prison-house, Virginia has no bolts or bars [that can stop me from sending you] my sympathy. In the name of the young girl sold from the warm clasp of a mother's arms to the clutches of a libertine or a profligate, in the name of the slave mother, her heart rocked to and fro by the agony of her mournful separations, I thank you, that you have been brave enough to reach out your hands to the crushed and blighted of my race. You have rocked the bloody Bastille; and I hope that from your sad fate great good may arise to the cause of freedom.... I would prefer to see Slavery go down peaceably by men breaking their sins by righteousness and their inequities by showing justice and mercy to the poor; but we cannot tell what the future may bring forth. God writes national judgments upon national sins; and what may be slumbering in the storehouse of divine justice we do not know. We may earnestly hope that your fate will not be a vain lesson, that it will intensify our hatred of Slavery and love of freedom, and that your martyr grave will be a sacred altar upon which men will record their vows of undying hatred to that system which tramples on man and bids defiance to God. I have written to your dear wife, and sent her a few dollars, and I pledge myself to you that I will continue to assist her....

Yours in the cause of freedom,
F. E. W.

African Americans who had escaped from slavery were often invited to tell their stories to gatherings of abolitionists. One man who was much in demand as a speaker was Henry "Box" Brown; he acquired his nickname by having friends ship him north in a packing crate. In 1849, Brown published the story of his escape.

I took my place in this narrow prison, with a mind full of uncertainty as to the result. It was a critical period of my life, I can assure you, reader; but if you have never been deprived of your liberty, as I was, you cannot realize the power of that hope of freedom....

I laid me down in my darkened home of three feet by two, and...resigned myself to my fate. My friend [sent] a telegraph message to his correspondent in Philadelphia, that such a box was on its way to his care....

With no access to the fresh air excepting three small...holes, I started on my perilous cruise. I was first carried to the express office, the box being placed on its end, so that I started with my head downwards, although the box was directed, "this side up with care." From the express office, I was carried to the depot...where I *happened* to fall "right side up," but...after a while...I was put aboard a steamboat, *and placed on my head.* In this dreadful position, I remained the space of an hour and a half...when I began to feel of my eyes and head, and found to my dismay, that my eyes were almost swollen out of their sockets, and the veins on my temple seemed ready to burst. I made no noise however, determining to obtain "*victory or death,*" but endured the terrible pain as well as I could, sustained...by the thoughts of sweet liberty.... One half hour longer and my sufferings would have ended in that fate, which I preferred to slavery; but I lifted up my heart to God in prayer...when to my joy, [two men] turned the box over.... One of these men in-

Henry "Box" Brown emerges from the crate that gave him his nickname. He became an eloquent spokesman for the abolitionist movement, describing first-hand the cruelties of slavery.

quired of the other, what he supposed that box contained, to which his comrade replied, that he guessed it was the mail. "Yes," thought I, "it is a *male,* indeed, although not the *mail* of the United States." ...

[The box eventually reached Philadelphia, where a wagon picked it up at the depot and took it] to the house of my friend's correspondent, where quite a number of persons were waiting to receive me. They appeared to be some afraid to open the box at first, but at length one of them rapped upon it, and with a trembling voice asked, "Is all right within?" to which I replied, "All right." The joy of these friends was excessive...each one seized hold of some tool, and commenced opening my grave. At length the cover was removed, and I arose, and shook myself...and I swooned away.

Frederick Douglass, born a slave about 1817, learned to read and write after his master's wife taught him the alphabet. Escaping in 1838, he began a career as an antislavery crusader.

Mattie J. Jackson, a former slave whose life story was printed in 1866, recalled her escape on the underground railroad.

I became acquainted with some persons who assisted slaves to escape by the underground railroad. They were colored people. I was to pretend going to church, and the man who was to assist and introduce me to the proper parties was to linger on the street opposite the house, and I was to follow at a short distance. On Sunday evening I begged leave to attend church, which was reluctantly granted if I completed all my work, which was no easy task. It appeared as if my mistress used every possible exertion to delay me.... Finally when I was ready to start, my mistress took a notion to go out to ride, and desired me to dress her little boy, and then get ready for church. Extensive hoops were then worn [under women's dresses], and as I had attached my whole wardrobe under mine by a cord around my waist, it required considerable dexterity and no small amount of maneuvering to hide the fact from my mistress. While attending to the child I had managed to stand in one corner of the room, for fear she might come in contact with me and thus discover that my hoops were not so elastic as they usually are.... I had nine pieces of clothing thus concealed on my person, and as the string which fastened them was small it caused me considerable discomfort. To my great satisfaction I at last passed into the street.... I saw my guide patiently awaiting me. I followed him at a distance until we arrived at the church, and there met two young ladies, one of whom handed me a pass and told me to follow them at a square's [a city block's] distance. It was now twilight. There was a company of soldiers about to take passage across the ferry, and I followed. I showed my pass, and proceeded up the stairs on the boat. While thus ascending the stairs the cord which held my bundle of clothing broke, and my feet became entangled in my wardrobe.... This was observed only by a few soldiers, who were too deeply engaged in their own affairs to interfere with mine. I seated myself in a remote corner of the boat, and in a few moments I landed on free soil for the first time in my life.

$100

REWARD

Ran away from the subscriber, living near the Anacostia Bridge, on or about the 17th November, negro girl ELIZA. She calls herself Eliza Coursy. She is of the ordinary size, from 18 to 20 years old, of a chestnut or copper color. Eliza has some scars about her face, has been hired in Washington, and has acquaintances in Georgetown.

I will give fifty dollars if taken in the District or Maryland, and one hundred dollars if taken in any free State; but in either case she must be secured in jail so that I get her again.

JOHN. P. WARING.

Nov. 28, 1857.

H. Polkinhorn's Steam Job Printing Office, D street, bet. 6th & 7th sts., Washington, D. C.

Owners of escaped slaves posted rewards for the return of their "property." After the passage of the Fugitive Slave Law, an escaped slave like Eliza was not safe even in the northern states.

EMANCIPATION

At top, ex-slave Hubbard Pryor, as he looked after escaping to a Union army camp during the Civil War. At bottom, what he looked like after enlisting in the Forty-Fourth Colored Infantry, an all-black unit of the Union army.

Many African Americans served in the Union Army during the Civil War. Lewis Douglass, son of the abolitionist Frederick Douglass, served in a Massachusetts regiment made up of northern free blacks and escaped slaves. He wrote to his future wife after a battle in 1863.

Morris Island, S.C. July 20
My Dear Amelia:

I have been in two fights, and am unhurt. I am about to go in another, I believe to-night. Our men fought well on both occasions. The last was desperate—we charged that terrible battery on Morris Island known as Fort Wagoner, and were repulsed with a loss of 3 killed and wounded. I escaped unhurt from amidst that perfect hail of shot and shell. It was terrible.... Should I fall in the next fight killed or wounded I hope to fall with my face to the foe. If I survive I shall write you a long letter....

This regiment has established its reputation as a fighting regiment—not a man flinched, though it was a trying time. Men fell all around me. A shell would explode and clear a space of twenty feet, our men would close up again, but it was no use we had to retreat, which was a very hazardous undertaking. How I got out of that fight alive I cannot tell, but I am here. My dear girl, I hope again to see you. I must bid you farewell should I be killed. Remember, if I die, I die in a good cause. I wish we had a hundred thousand colored troops—we would put an end to this war. Good Bye to all. Write soon.

Your own loving
Lewis

As the Union soldiers approached their plantations, many slaves decided to escape to the safety of the Union forces. Mary Barbour of North Carolina told her story in the 1930s at the age of 81.

One of the first things that I remembers was my pappy waking me up in the middle of the night, dressing me in the dark, all the time telling me to keep quiet.

After we dressed, he went outside and peeped around for a minute, then he comed back and got us. We snook out of the house and along the woods path, Pappy toting one of the twins and holding me by the hand and Mammy carrying the other two.

I reckons that I will always remember that walk, with the bushes slapping my legs, the wind sighing in the trees, and the hoot owls and whippoorwills hollering at each other from the big trees. I was half asleep and scared stiff, but in a little while we pass the plum thicket and there am the mules and wagon.

There am the quilt in the bottom of the wagon, and on this they lays we younguns. And Pappy and Mammy gets on the board across the front and drives off down the road....

We travels all night and hid in the woods all day for a long time, but after awhile [they reached the camp of the Union army]....

My Pappy was a shoemaker, so he makes Yankee boots, and we gets along pretty good.

Harriet Tubman, the legendary African American conductor on the Underground Railroad, went to the battlefields with the First South Carolina Volunteers, a Union army regiment made up primarily of escaped slaves. She described seeing the slaves who came in from the countryside, seeking freedom.

I never saw such a sight. We laughed, and laughed, and laughed. Here you'd see a woman with a pail on her head, rice smoking in it just as if she'd taken it from the fire, young one hanging on behind, one hand hanging around her forehead to hold on, another hand digging into the rice pot, eating with all its might; holding on to her dress two or three more; down her back a bag with a pig in it. One woman brought two pigs, a white one and a black one; we took them all aboard; named the white pig Beauregard [a Confederate general], and the black pig Jeff Davis [president of the Confederacy]. Sometimes the women would come with twins hanging around their necks; appears like I never seen so many twins in my life; bags on their shoulders, baskets on their heads, and young ones tagging behind, all loaded; pigs squealing, chickens screaming, young ones squalling.

Miles Mark Fisher's son Elijah wrote of the day in 1863 when the Yankee soldiers arrived on the plantation where Miles was a slave. Miles's master, Doctor Ridley, read the message that the soldiers brought and told Fisher to summon the other slaves.

As they left the field [the slaves] wondered who was to be whipped or who was to be sold or what orders were to be given. Half-startled, half-afraid, they wended their way through the fields in one silent mass of praying creatures....

The master was weeping bitterly.... His words were stifled with sobs. The slaves were awe-stricken; they had never seen a white man cry. Only slaves had tears, they thought. All eyes were fastened on Doctor Ridley. He was saying something. "All you niggers—all you niggers are free as I am." The surprise was shocking, but in an instant in his usual harsh voice he added: "But there ain't going to be any rejoicing here. Stay here until the crop is made, and I'll give you provisions. Go back to work."

But the slaves did rejoice and loudly, too. Some cried; some jumped up and cracked their heels. Charlotte [Miles's wife] took her younger children in her arms and shouted all over the plantation: "Chillun, didn't I tell you God 'ould answer prayer?"

From the outbreak of the Civil War, free African Americans were eager to fight for the freedom of their brothers and sisters in bondage. The following letter was sent to Secretary of War Simon Cameron:

Cleveland, O., 15 Nov. 1861.
Sir:

We would humbly and respectfully State, that we are Colard men (legal voters) all voted for the present administration. The question now is will you allow us the poor priverlige of fighting—and if need be dieing—to suport those in office who are own choise. We believe that a reigement of colard men can be raised in this State, who we are sure, would make as patriotic and good Soldiers as any other.

What we ask of you is that you give us the proper athroity to rais such a reigement, and it can and SHALL be done.

We could give you a Thousand names as ether signers or as refferance if required....

W. T. Boyd
J. T. Alston
P.S. we waite your reply.

The first black units of the Union army were formed in July 1862. Initially composed of already-free blacks, their ranks swelled with ex-slave volunteers. A total of 178,895 African Americans eventually served. Black women also aided the cause as military nurses.

The Cheatham family in Virginia during Reconstruction. During the 12-year period when the federal government protected their rights as citizens, some southern blacks were able to start their own farms and businesses.

CHAPTER FOUR

A NEW LIFE

At the end of the Civil War, there were about 4 million newly freed African Americans in the slave states. Virtually all were illiterate and without money to start a new life. For the next 12 years, a period known as Reconstruction, the nation struggled with the problem of how to find a place for them in American society. The 14th Amendment, which was ratified in 1868, made the former slaves citizens and guaranteed their constitutional rights. Two years later, the 15th Amendment affirmed the right of black men to vote. Indeed, during Reconstruction 16 African Americans from the South were elected to the House of Representatives and two others became U.S. senators.

But giving the slaves legal rights did not guarantee economic freedom for them. Through their labor, the slaves had made fortunes for their owners. Some Americans, both white and black, felt that the nation owed the former slaves a debt. In 1862, Representative Thaddeus Stevens introduced a bill in Congress to confiscate slave owners' land and distribute it to slaves. In the slave cabins of the South, rumors spread that each family was to be given "40 acres and a mule" to start their own farms.

During the Civil War, Union General William T. Sherman did in fact seize plantations in Georgia and South Carolina. He distributed 40-acre plots of land to about 40,000 ex-slave families in these states. However, President Andrew Johnson, a native of Tennessee, decided that reconciliation between North and South was more important than redressing the wrongs of slavery. Johnson ordered the army to remove the black settlers from the land Sherman had given them.

The Republican-controlled Congress opposed Johnson's policies (and nearly impeached him). In 1866 Congress passed the Southern Homestead Act, which opened government-owned lands in five southern states to African American settlers, allowing them to pay for the land in installments.

Congress also created the Freedmen's Bureau to provide legal, health, and educational services to the ex-slaves. Black and white teachers and nurses from the North distributed food and clothing and established schools and hospitals. Officials of the Freedmen's Bureau intervened on behalf of blacks in state courts, registered them to vote, and tried to reunite families that had been separated at slave sales. Members of the Freedmen's Bureau helped to establish Fisk University in Nashville, Tennessee, and Howard University in Washington, D.C., the first African American institutions of higher learning.

During Reconstruction, some blacks in the South started their own businesses, such as restaurants, stores, and barbershops. A few worked in larger operations, including railroads and steamship lines. A number of blacks formed cooperatives in which they pooled their funds and managed to borrow enough money to buy land or equipment.

By 1900, 25 percent of black farmers in the South owned their own farms. A few were actually wealthy, such as U.S. Senator Blanche K. Bruce, who owned a thousand-acre plantation in Mississippi. But the greatest number of black landowners were outside the Deep South. In Virginia, 60 percent of the black farmers owned their own land, but in Georgia the figure was only 14 percent and in Mississippi, 11 percent.

In many instances, blacks continued to work for their old masters after emancipation, sometimes under conditions that were little better than slavery. White southern planters devised ways to keep the ex-slaves in economic bondage. (One South Carolina planter even made his workers sign a contract that required them to call him "master.")

Tenant farming was the most common system. The landowners allocated plots of land to black families, usually providing them with seed, tools, animals, and shelter. In return, the tenants had to pay an annual rent from the sale of their crop. Some tenant farmers, called sharecroppers, agreed to give the landowner a percentage of their crop, usually one-half to two-thirds. Whatever profit a tenant family could get from the rest of the crop might eventually be saved to buy their own land.

Some tenant farmers were able to become independent. The majority, however, were cheated by being made to sign contracts they could not read or understand. Many often had to borrow money at high rates of interest, hoping that a good crop would enable them to pay back the debt. When cotton prices dropped or bad weather resulted in a poor crop, they were forced to borrow even more to carry them over to the next year.

When bankers foreclosed on the debt, black farm families fell to a lower level—called "debt peonage." The banks forced them to work off their debt, selling their services to plantation owners. Such workers were compelled by law to remain on the plantations until they paid back what they owed. Plantation owners also leased the labor of convicts, some of whom had been jailed on petty or trumped-up charges merely to provide cheap labor.

There were fewer ways to escape from this system of economic bondage than there had been ways to escape slavery. After the Civil War, opportunities for African Americans in the North declined. In part, this situation resulted from a great increase in white immigrants from Germany, Ireland, and other European nations. Very often in the 19th century, Irish and blacks in northern cities competed for the same unskilled jobs. During the Civil War, Irish in New York rioted to protest the military draft, killing African Americans and burning their homes. Other immigrants took over some of the skilled trades that blacks had formerly held in both the North and South.

A black family poses proudly in their carriage. Prosperous blacks like these were often targets of racist groups.

Some former slaves looked to the undeveloped regions of the West for opportunity. From 1800, fur traders and hunters, some of them of African descent, had explored as far west as Oregon. Several thousand blacks had been among the prospectors in the California gold rush of 1849. By 1856, San Francisco's African American population was large enough to support its own churches, schools, and a newspaper. In the postwar years, African Americans in the U.S. Army, known as "Buffalo Soldiers," took part in the wars against some of the Native American tribes in the West. Others found work as miners in Colorado, Nevada, and Utah.

In 1879, Benjamin "Pap" Singleton, a former Tennessee slave, organized a movement of African Americans into Kansas. Claiming government land under the Homestead Act of 1862, these ex-slaves sometimes encountered racial prejudice from white settlers, but many black homesteaders succeeded. Some of them began to move into Oklahoma, which had earlier been set aside for Native Americans. Indeed, the Cherokees and other tribes that had been forced off their land in Georgia had owned slaves and sometimes intermarried with blacks. Many all-black towns were founded in the Oklahoma Territory, and there was a short-lived movement to make it an all-black state.

Reconstruction ended in 1877 after Rutherford B. Hayes won a disputed Presidential election with the crucial support of the southern states. Hayes withdrew the last of the federal troops that occupied the South. The nation turned its back on the former slaves, leaving them to the mercies of their old masters.

Gradually, the southern states passed laws to keep black men from voting. Deprived of political power, the African Americans had no voice in local or state governments. Racist groups such as the Ku Klux Klan terrorized black families who were too prosperous or showed insufficient respect to whites. Between 1882 and 1900, about 3,000 southern blacks were lynched and burned to death. Many more were dragged from their homes, flogged, and tortured.

To reinforce the idea of white supremacy, the races were systematically separated in all areas of life. Blacks could not eat with whites, attend school with them, or use the same bathrooms and drinking fountains. Theaters, railway cars, and even courthouses had separate sections for blacks.

The Supreme Court disgraced itself by approving this "Jim Crow" system (named for a song mocking blacks that was popularized by a white entertainer). In 1883, the Court struck down the 1875 Civil Rights Act, which said public accommodations must be open to all citizens. In the case of *Plessy* v. *Ferguson* (1896), the Supreme Court upheld a Louisiana law requiring railroads to provide "equal but separate accommodations for the white and colored races." The *Plessy* decision was used to justify the establishment of a separate school system for blacks (but it was hardly equal). President Woodrow Wilson, a Virginian, reinforced the system by segregating public facilities in federal office buildings.

By the end of the 19th century, African Americans faced prejudice throughout the nation. With little political or economic power, forced into menial jobs and near-slavery, many of them seemed resigned to a permanent condition of poverty and humiliation.

At the beginning of the 20th century, two African American leaders expressed different ideas about the course blacks should follow. Booker T. Washington, the head of Tuskegee Institute, advised them to accept segregation and "social inequality." He felt that with the help of sympathetic whites, African Americans could receive vocational training that would allow them to move up the economic ladder. A more militant view came from W. E. B. Du Bois, a black educator who had earned a doctorate at Harvard University. Du Bois opposed all forms of discrimination and believed that well-educated blacks should lead the fight to demand full rights for all African Americans.

This black-owned store in Virginia celebrates Emancipation Day in 1888 by displaying a banner with Abraham Lincoln's picture.

The opposing views of these two men would be debated by their followers for decades. Yet Robert S. Abbott, the editor of the *Chicago Defender,* a black newspaper, possibly had a greater direct effect on the lives of the majority of African Americans. Abbott's newspaper circulated widely among southern blacks, who passed it from hand to hand. He continually urged them to abandon their lives of poverty and oppression and move to the North.

Abbott's message set off what became known as the Great Migration. Between 1910 and 1950, about 5.6 million African Americans left the tenant farms and towns of the South to settle in northern cities. More followed until about 1970. This event was one of the greatest population shifts the United States has ever experienced. In the process, African Americans brought with them the blues, jazz, and gospel music that were to transform the culture of America.

Many migrants took the railroad from the Deep South to the Illinois Central Station at 12th Street and Michigan Avenue in Chicago. Carrying cardboard suitcases and dreams of a better life, millions passed through what has been called the Ellis Island of African Americans. The black population of Chicago alone grew from 44,000 to 234,000 between 1910 and 1930.

The first migration of African Americans—from their homeland during the centuries of the slave trade—had been an involuntary one. But the Great Migration was one African Americans chose to make themselves. They moved for the same reasons that brought other immigrant groups to the United States—to escape persecution and injustice and to find a better life.

Without the money to purchase their own land, many blacks were forced to work as laborers or sharecroppers in a system that was not much better than slavery.

FARMING IN THE SOUTH

African Americans never got the "40 acres and a mule" that they had hoped the government would give them to make a new start. But some managed to acquire land of their own. In the 1930s, 94-year-old Josephine Smith of North Carolina recalled her struggle to buy a farm.

Right after the war...some sort of corporation cut the land up, but the slaves ain't got none of it that I ever heard about. I got married before the war to Joshua Curtis. Josh ain't really care about no home, but through this land corporation I buyed up these fifteen acres on time [payments]. I cut down the big trees that was all over these fields, and I mauled out the wood and sold it, then I plowed up the fields and planted them. Josh did help to build the house, and he worked out [for other employers] some.

I done a heap of work at night too, all of my sewing and such.... I finally paid for the land. Some of my chilluns was born in the field, too.... I blowed in a bottle to make the labor quick and easy. All of this time I had nineteen chilluns, and Josh died, but I kept on, and the fifteen what is dead lived to be near about grown, every one of them.

I'll never forget my first bale of cotton and how I got it sold. I was some proud of that bale of cotton, and after I had it ginned, I set out with it on my steer cart for Raleigh. The white folks hated the nigger then, specially the nigger what was making something, so I dasn't ask nobody where the market was. I thought I could find the place by myself, but I rid all day and had to take my cotton home with me that night, 'cause I can't find no place to sell it at. But that night I think it over, and the next I go back and ask a policeman about the market. Lo and behold, child, I found it on Blount Street, and I had pass by it several times the day before.

Southern whites often doubted that the ex-slaves could survive without the support of their masters. Nellie A. Plummer, the daughter of former slaves, described how her family worked to prove the doubters wrong.

On July 14, 1868, father bound the bargain with B.F. Guy [a white landowner] for a hill adjoining Riverdale, containing ten acres, more or less, for the sum of $1,000 by paying him $344.75. This meant deprivation such as you, of this day and time, know not of—almost starvation....

By September 26, 1868, Guy sent for another payment (as if money grew on bushes for the freed men). That evening [father] carried him $160.25, making $505 paid! Hard? worse than that, but the thought of being in our own home urged them on! Father, mother, sister, Henry, Julia, and Saunders worked out [to other employers] and gave all they could make. By January 17, 1870, father had paid the entire thousand dollars! Much to Guy's surprise. For he was a speculator.

He never dreamed that father would or could pay for it in the specified time—two years! So when it was completed in 18 months, it was indeed a wonder! Guy's neighbors had said to him: "You are ruining our country!"

"How is that," said Guy.

"Why [by] selling Negroes land."

Guy would reply: "Don't worry, they can't raise the money. In time, I'll take the land back."

But he didn't know the man with whom he was dealing!... By September, 1870, father had finished building our four-room log house, and we moved...into a happier place—Our Own Home!

But many African Americans in the South found their situation barely improved after the Civil War. They worked under conditions that were akin to slavery. Such was the story of a Georgia man.

Shortly after the [Civil] war my mother died, and I was left to the care of my uncle.... When I was about ten years old my uncle hired me out to Captain——. I had already learned how to plow, and was also a good hand at picking cotton. I was told that the Captain wanted me for his house-boy, and that later on he was going to train me to be his

A woman carries crops from her garden to market. Until 1910, a majority of African Americans earned their living in agriculture.

Cotton remained the primary crop in most of the South until about 1915. Freedom did little to change the lives of these cotton-field harvesters, whose wages barely provided enough money to survive.

For many African Americans after the Civil War, tenant farming was just a step above slavery. This contract written in 1881 spelled out the terms for tenants on the Grimes plantation in North Carolina:

To every one applying to rent upon shares, the following conditions must be read and agreed to.

To every 30 or 35 acres, I agree to furnish the team, plow, and farming implements.... The croppers are to have half of the cotton, corn and fodder (and peas and pumpkins and potatoes if any are planted) if the following conditions are complied with, but--if not—they are to have only two-fifths. Croppers are to have no part of interest in the cotton seed raised from the crop planted and worked by them. No vine crops of any description, that is no watermelons...squashes or anything of that kind...are to [be] planted in the cotton or corn. All must work under my direction....

All croppers must clean out stables and fill them with straw, and haul straw in front of stables whenever I direct. All the cotton must be manured...the croppers to pay for one half of all manure bought, the quantity to be purchased for each crop must be left to me.

No cropper to work off the plantation when there is any work to be done on the land he has rented, or when his work is needed by me or other croppers....

Every cropper must be responsible for all gear and farming implements placed in his hands, and if not returned must be paid for unless it is worn out by use.

Croppers must sow and plow in oats and haul them to the crib, but must have no part of them. Nothing to be sold from their crops, nor fodder, nor corn to be carried out of the fields until my rent is all paid, and all amounts they owe me and for which I am responsible are paid in full....

The sale of every cropper's part of the cotton to be made by me when and where I choose to sell, and after deducting all they may owe me and all sums that I may be responsible for on their accounts, to pay them their half of the net proceeds. Work of every description, particularly the work on fences and ditches, to be done to my satisfaction, and must be done over until I am satisfied that it is done as it should be.

coachman. To be a coachman in those days was considered a post of honor, and as young as I was, I was glad of the chance. But I had not been at the Captain's a month before I was put to work on the farm, with some twenty or thirty other negroes—men, women and children. From the beginning the boys had the same tasks as the men and women. There was no difference. We all worked hard during the week, and would frolic on Saturday nights and often on Sundays. And everybody was happy. The men got $3 a week and the women $2. I don't know what the children got. Every week my uncle collected my money for me, but it was very little of it that I ever saw. My uncle fed and clothed me, gave me a place to sleep, and allowed me 10¢ or 15¢ a week for "spending change," as he called it. I must have been 17 or 18 years old before I got tired of that arrangement, and felt that I was man enough to be working for myself and handling my own wages.... Unknown to my uncle or the Captain I went off to a neighboring plantation and hired myself out to another man. The new landlord agreed to give me 40¢ a day and furnish me one meal. I thought that was doing fine. Bright and early one Monday morning I started for work, still not letting the others know anything about it. But they found it out before sundown. The Captain came over to the new place and brought some kind of officer of the law. The officer pulled out a long piece of paper from his pocket and read it to my new employer. When this was done I hear my new boss say:

"I beg your pardon, Captain. I didn't know this nigger was bound out to you, or I wouldn't have hired him."

"He certainly is bound out to me," said the Captain. "He belong to me until he is twenty-one, and I'm going to make him know his place."

So I was carried back to the Captain's. That night he made me strip off my clothing down to my waist, had me tied to a tree in his backyard, ordered his foreman to give me thirty lashes with a buggy whip across my bare back, and stood by until it was done. After that experience the Captain made me stay on his place night and day,—but my uncle still continued to "draw" my money.

Few tenant farmers found it possible to get ahead. An African American in Mississippi described his plight in the 1870s.

Just before picking time comes, the merchant sends men around from place to place to see how the crop is getting along; how many bales will probably be made. By this the merchant knows how large to make his bill.... If the colored man refuses to pay the bill, which, as I have said, is always made large enough to cover the value of the entire crop, after paying the rent, the merchant comes into court and sues him. The white man brings his itemized account into court; the colored man has no account, and of course he is beaten in the suit, and the cost is thrown onto him. They stand against him, if he cannot pay it. And colored men soon learn that it is better to pay any account, however unjust, than to refuse, for he stands no possible chance of getting justice before the law.

Many African Americans worked on long-term contracts that were strictly enforced by state officials. One man who worked in a Georgia camp for three years to pay off his debts described conditions to an interviewer at the turn of the 20th century.

The next morning it was explained to us by the two guards appointed to watch us that, in the papers we had signed the day before, we had not only made acknowledgement of our indebtedness, but that we had also agreed to work for the Senator [the camp owner] until the debts were paid by hard labor. And from that day forward we were treated just like convicts. Really we had made ourselves lifetime slaves, or peons, as the laws called us. But, call it slavery, peonage, or what not, the truth is we lived in a hell on earth what time we spent in the Senator's peon camp....

The stockades in which we slept were, I believe, the filthiest places in the world. They were cesspools of nastiness.... I am willing to swear that a mattress was never moved after it had been brought there, except to turn it over once or twice a month. No sheets were used, only dark-colored blankets. Most of the men slept every night in the clothing that they had worked in all day.... The doors were locked and barred each night, and tallow candles were the only lights allowed. Really the stockades were but little more than cow sheds, horse stables or hog pens....

The working day of a peon farm begins with sunrise and ends when the sun goes down; or in other words, the average people works from ten to twelve hours each day, with one hour (from 12 o'clock to 1 o'clock) for dinner. Hot or cold, sun or rain, this is the rule. As to their meals.... Each peon is provided with a great big tin cup, a flat tin pan and two big tin spoons.... At meal time the peons pass in single file before the cooks, and hold out their pans and cups to receive their allowances. Cow peas (red or white, which when boiled turn black), fat bacon and old-fashioned Georgia corn bread,...from one to two and three inches thick, make up the chief articles of food....

But I didn't tell you how I got out. I didn't get out—they put me out. When I served as a peon for nearly three years...one of the bosses came to me and said that my time was up.... He gave me a new suit of overalls, which cost 75¢, took me in a buggy and carried me across the Broad River into South Carolina, set me down and told me to "git." I didn't have a cent of money, and I wasn't feeling well, but somehow I managed to get a move on me. I begged my way to Columbia. In two or three days I ran across a man looking for laborers to carry to Birmingham, and I joined his gang. I have been here in the Birmingham district since they released me, and I reckon I'll die either in a coal mine or an iron furnace. It don't make much difference which. Either is better than a Georgia peon camp. And a Georgia peon camp is hell itself!

As late as 1909, more than half of the cotton produced in the United States was grown by African American farmers. This young man brings a few bales into Camden, South Carolina, around 1900.

Rice, the crop that this couple is tending, requires intensive labor. The husband pulls a plow guided by his wife to cultivate the field.

PIONEERING IN THE WEST

Though the sheriff in movies and TV shows is always white, African Americans wore badges in tough western towns such as Lawrence, Kansas, where this man kept the peace.

James Beckwourth was born around 1798, the son of a white Revolutionary War veteran and an African American slave. After the United States bought the vast Louisiana Territory, James's family moved to a farm near St. Louis. His white father, who kept his children enslaved, apprenticed James to a blacksmith. Soon afterward James ran away. He joined an expedition that was heading farther west to collect furs.

James Beckwourth became one of the storied "mountain men" who explored the uncharted areas of the American West. At one point, he found refuge with the Crow tribe, and became one of their chiefs. In 1854, he dictated a memoir of his fabulous career.

In the following passage, Beckwourth describes how he discovered the pass over the Sierra Nevada Mountains that bears his name. Having arrived in California during the gold rush, Beckwourth and a group of companions traveled eastward from the American Valley, the region where the first great gold strikes had been made.

It was the latter end of April when we entered upon an extensive valley.... Swarms of wild geese and ducks were swimming on the surface of the cool crystal stream, which was the central fork of the Rio de las Plumas, or sailed the air in clouds over our heads. Deer and antelope filled the plains, and their boldness was conclusive that the hunter's rifle was to them unknown.... We struck across this beautiful valley to the waters of the Yuba, from thence to the waters of the Trucy.... This I at once saw, would afford the best waggon-road into the American Valley approaching from the eastward. [Some of his companions] thought highly of the discovery, and even proposed to associate with me in opening the road.... [This could prove more profitable than looking for gold, for a reliable overland route did not then exist between California and the eastern United States.]

A homesteading family in the Oklahoma Territory in 1889. Many African American "exodusters," who left the South to seek a better life in the West, went to Oklahoma.

When I reached Bidwell's Bar and unfolded my project, the town was seized with a perfect mania for the opening of the route. The subscriptions toward the fund required for its accomplishment amounted to five hundred dollars.... While thus busily engaged, I was seized with erysipelas [a skin disease], and abandoned all hopes of recovery.... I made my will, and resigned myself to death. Life still lingered in me, however, and a train of waggons came up, and encamped near to where I lay.... [Beckwourth told the wagon-drivers of his discovery.] They offered to attempt the new road if I thought myself sufficiently strong to guide them through it. The women, God bless them! came to my assistance, and through their kind attentions and excellent nursing I rapidly recovered from my lingering sickness, until I was soon able to mount my horse, and lead the first train, consisting of seventeen waggons, through "Beckwourth's Pass."

[Beckwourth established a ranch and trading-post at the pass.] My house is considered the emigrant's landing-place, as it is the first ranch he arrives at in the golden state, and is the only house between this point and Salt Lake.

In 1879, many African Americans in Louisiana, Mississippi, Texas, and Tennessee found conditions unbearable in the South and decided to head for undeveloped areas farther west. John Solomon Lewis took his family from Louisiana to Kansas.

You see I was in debt, and the man I rented land from said every year I must rent again to pay the other year, and so I rents and rents, and each year I gets deeper and deeper in debt. In a fit of madness I one day said to the man I rented from: "It's no use, I works hard and raised big crops and you sells it and keeps the money, and brings me more and more in debt, so I will go somewhere else and try to make headway like white workingmen."

He got very mad and said to me: "If you try that job, you will get your head shot away."

So I told my wife, and she says: "Let us take to the woods in the night time." Well we took [to] the woods, my wife and four children, and we was three weeks living in the woods waiting for a boat. Then a great many more black people came and we was all together at the landing.

Boats came along, but they would not stop, but before long the *Grand Tower* hove up and we got on board. Says the captain, "Where's you going?"

Says I, "Kansas."

Says he, "You can't go on this boat."

Says I, "I do; you know who I am. I am a man who was a United States soldier and I know my rights, and if I and my family gets put off, I will go in the United States Court and sue for damages."

Says the captain to another boat officer, "Better take that nigger or he will make trouble."...

When I landed on the soil, I looked on the ground and I says this is free ground. Then I looked on the heavens, and I says them is free and beautiful heavens. Then I looked within my heart, and I says to myself I wonder why I never was free before?

When I knew I had all my family in a free land, I said let us hold a little prayer meeting; so we held a little meeting on the river bank. It was raining but the drops fell from heaven on a free family, and the meeting was just as good as sunshine. We was thankful to God for ourselves and we prayed for those who could not come.

Benjamin Singleton

Benjamin "Pap" Singleton grew up in slavery, serving as a cabinetmaker in Nashville. After being sold to owners in Mississippi, he escaped and went to Canada. Soon he was in Detroit, where he ran a boardinghouse that sheltered fugitive slaves. After the Civil War, he returned to Nashville, where he devoted his efforts to the needs of the newly freed slaves. When he saw that conditions in the South were getting worse rather than better, Singleton, then 70 years old, became a leader in bringing people to Kansas. In 1880, he explained how the Kansas trek had begun:

"My people, for the want of land—we needed land for our children—and their disadvantages—that caused my heart to grieve and sorrow; pity for my race, sir, that was coming down, instead of going up—that caused me to go to work for them. I sent out [to Kansas] perhaps in '66—perhaps so; or in '65, any way—my memory don't recollect which; and they brought back tolerable favorable reports; then I jacked up three or four hundred and went into Southern Kansas, and found it was good country, and I thought Southern Kansas was congenial to our nature, sir; and I formed a colony there, and bought about a thousand acres of ground—the colony did—my people."

Arthur Walker, one of the many black cowboys of the western frontier. Such men took part in the cattle drives that brought the herds overland from Texas ranches to the railroads farther north.

Oklahoma was another destination for the former slaves. Mabel Murphy told the story of her father's journey.

My father was twelve years old when freedom was declared.... He married a Cherokee Indian, so my mother was an Indian. He never allowed us to say we were afraid of anything. He'd always tell us, "No, you have no need to be afraid of nothing. You show fear, you are defeated. If you stand up and be brave, you can master it." We used to live in Oklahoma where the Indians would come through sometimes on their horses. Sometimes we'd want to be afraid of them. He'd tell us, "No, don't be afraid. Now you just watch when they come by this time." So my dad wouldn't take no gun or no weapon or nothing. He'd just walk out and offer them rest. They'd fall off their horses. They'd say, "Hm, shake hand." They never did offer to fight my dad. We lived in the early part of the Indian fights in Oklahoma. We never were attacked by them. They'd come by and a lot of times they would want food. My mother would feed them; then they'd go on about their business. They never did harm us, because my mother was Cherokee anyway. Her hair was down to her waist, and when they took one look at her they knew she was Indian. My father always treated them kind. They never done us any harm.

Many African Americans were among the cowhands of the Old West who tended and drove the cattle herds. One of them, a former slave named Nat Love, wrote his autobiography in 1907. He described how he won the nickname "Deadwood Dick" on the Fourth of July, 1876, in the town of Deadwood in the Dakota Territory.

Our trail boss was chosen to pick out the mustangs from a herd of wild horses just off the range, and he picked out twelve of the most wild and vicious horses that he could find.

The conditions of the contest were that each of us...was to rope, throw, tie, bridle and saddle, and mount the particular horse picked for us in the shortest time possible....

It seems to me that the horse chosen for me was the most vicious of the lot. Everything being in readiness, the "45" cracked and we all sprang forward together, each of us making for our particular mustang.

I roped, threw, tied, bridled, saddled, and mounted my mustang in exactly nine minutes from the crack of the gun. The time of the next nearest competitor was twelve minutes and thirty seconds. This gave me the record and championship of the West, which I held up to the time I quit the business in 1890, and my record has never been beaten. It is worthy of passing remark that I never had a horse pitch with me so much as that mustang, but I never stopped sticking my spurs in him and using my quirt on his flanks until I proved his master. Right there the assembled crowd named me Deadwood Dick and proclaimed me champion roper of the western cattle country.

Era Bell Thompson, the granddaughter of slaves, came to North Dakota with her family in 1914. They settled in the town of Driscoll. Thompson described a Christmas spent with one of the few other black families in the state.

By eight o'clock, Sport [the dog] and the whole family were bundled up in blankets and bedded down in the straw in the big sled. The farther we went, the bigger the farms, the farther apart. Rising up out of the snow were large, prosperous houses and barns, many of them flanked by sturdy groves of cottonwoods, straight, symmetrical rows that served as windbreaks. Trees! We could hardly believe it....

It was late afternoon when we came in sight of the Williams farm. Sport leaped over the side of the sled and joined the barking dogs that came running down the road to meet us.

"Merry Christmas, and welcome!" bellowed Mack Williams, as we pulled up in front of his door.

"Merry Christmas to you, sir." Pop climbed out of the sled and gave him the Odd Fellows' grip. [The Odd Fellows were a fraternal organization.]

"Brother Thompson. Bro—ther Thompson!" Mack jumped up and down, his big body shaking, smiles wreathing his dark face.

"You must be starved," said Mrs. Williams. "Dinner's ready. We've been waiting for two hours. Afraid you'd got lost."

Circulation slowly returned to our numbed limbs. Under the warmness of welcome, our tongues had loosened, and soon everybody was talking, laughing, shouting like old friends. Mrs. Williams hugged Mother, and somebody kissed me squarely in the mouth before I could fight my way out of the wool muffler.

Half an hour later we were seated at a table groaning with food. We ate and talked, ate and rested, then started all over again, as an unending stream of food flowed from kitchen to table. Food was urged upon us, pushed off on our plates against first mild, then vigorous, protests. Near the end of the meal we sat back dawdling with the food. Slumping down with the weight of it, I mournfully watched the mound of ice cream melt before my very eyes, melt faster than I could eat. No one scolded for half-empty plates. Mack raised his mighty voice in laughter as, one by one, we children grudgingly withdrew.

"I like to see the young 'uns eat," he said. "Look at my brood; fat as hogs every one of 'em. Fat, black and sassy." Mack Williams was proud of his blackness.

While the grown-ups were all sitting around the table, Ted Williams, Mack's brother, and his family arrived.... Now there were fifteen of us, four percent of the state's entire Negro population. Out there in the middle of nowhere, laughing and talking and thanking God for this new world of freedom and opportunity, there was a feeling of brotherhood, of race consciousness, and of family solidarity.... I was part of a whole family, and my family was a large part of a little colored world, and for a while no one else existed.

"Doc" Hisom, a pioneer who built his cabin in Melba, Idaho, on the Snake River.

The bravery shown by African American troops during the Civil War persuaded Congress to make them a permanent part of the army. Most of the segregated black units in the 19th century were assigned to fight Native Americans, who dubbed them "buffalo soldiers."

JIM CROW

"Jump Jim Crow" is said to have been the name of a song used by a disabled black entertainer. A white performer (shown here) popularized it in his minstrel act, in which he darkened his face with makeup and ridiculed blacks. Over time, the name Jim Crow *was applied to the system of segregation and discrimination enforced by southern state laws.*

White southerners founded the Ku Klux Klan in 1866 to terrorize newly freed African Americans. Klan members dragged people from their homes to hang, burn, and flog them. In 1871, Betsey Westbrook testified on the steps of the courthouse in Demopolis, Alabama, about the killing of her husband, Robin Westbrook. His "crime" was his open support of the Radical Republicans who controlled Congress.

They came up behind the house. One of them had his face smutted and another had a knit cap on his face. They first shot about seven [shotgun] barrels through the window. One of them said, "Get a rail and bust the door down." They broke down the outside door.... One of them said, "Raise a light."... Then they saw where we stood and one of them says, "You are that damned son of a bitch Westbrook." The man had a gun and struck him on the head. Then my husband took the dog-iron and struck three or four of them. They got him jammed up in the corner and one man went around behind him and put two loads of a double-barreled gun in his shoulders. Another man says, "Kill him, God damn him," and took a pistol and shot him down. He didn't live more than half an hour.

My boy was in there while they were killing my husband and he says, "Mammy, what must I do?" I says, "Jump outdoors and run." He went to the door and a white man took him by the arm and says, "God damn you, I will fix you too," but he snatched himself loose and got away.

As an old man, W. L. Bost described the terrors of the Ku Klux Klan in North Carolina.

Then the Ku Klux Klan come along. They were terrible dangerous. They wear long gowns, touch the ground. They ride horses through the town at night, and if they find a Negro that tries to get nervy or have a little bit for himself, they lash him nearly to death and gag him and leave him to do the best he can. Sometimes they put sticks in the top of the tall thing [pointed hood] they wear and then put an extra head up there with scary eyes and great big mouth, then they stick it clear up in the air to scare the poor Negroes to death.

They had another thing they call the Donkey Devil that was just as bad. They take the skin of a donkey and get inside of it and run after the poor Negroes. Oh, Miss, them was bad times, them was bad times. I know folks think the books tell the truth, but they sure don't. Us poor niggers had to take it all.

There were more than 3,400 recorded lynchings of African Americans between 1882 and 1968. In 1900 and 1922, Congress defeated two proposals to make lynching a federal crime.

By the early years of the 20th century, the Jim Crow laws of segregated public facilities had spread throughout the South. An African American woman whose father had been a slave explained to interviewers in 1904 the humiliation of the system.

Not long since I visited a Southern city where the "Jim Crow" car law is enforced. I did not know of this law, and on boarding an electric car took the most convenient seat. The conductor yelled, "What do you mean? Niggers don't sit with white folks down here. You must have come from 'way up yonder. I'm not Roosevelt. We don't sit with niggers, much less eat with them."

I was astonished and said, "I am a stranger and did not know of your law." His answer was: "Well, no back talk now; that's what I'm here for—to tell niggers their places when they don't know them."

Every white man, woman and child was in a titter of laughter by this time at what they considered the conductor's wit.

These fine Southern men and women, who pride themselves on their fine sense of feeling, had no feeling for my embarrassment and unmerited insult, and when I asked the conductor to stop the car that I might get off, one woman said in a loud voice, "These niggers get more impudent every day; she doesn't want to sit where she belongs."

No one of them thought that I was embarrassed, wounded and outraged by the loud, brutal talk of the conductor and the sneering, contemptuous expressions on their own faces. They considered me "impudent" when I only wanted to be alone that I might conquer my emotion. I was nervous and blinded by tears of mortification which will account for my second insult on this same day.

I walked downtown to attend to some business and had [to] take an elevator in an office building. I stood waiting for the elevator, and when the others, all of whom were white, got in I made a move to go in also, and the boy shut the cage door in my face. I thought the elevator was too crowded and waited;

Among the everyday humiliations in a segregated society were separate public drinking fountains. Blacks could not sit with whites in theaters, restaurants, or waiting rooms. In 1913, President Woodrow Wilson extended the system of segregation to federal office buildings.

The Ku Klux Klan was founded by a former Confederate general in 1866. Hiding under their white robes, Klan members sought to terrorize black Americans and keep them subservient to whites.

Booker T. Washington

In 1895, a new African American leader came to national attention with a widely quoted speech in Atlanta. He was Booker T. Washington, born a slave in 1856 and later the head of Tuskegee Institute, a prestigious African American school in Alabama. In what became known as the "Atlanta Compromise," Washington advised his fellow African Americans not to seek "social equality," an end to segregation, or even the right to vote. Instead, they should rely on self-help:

"Cast down your bucket where you are.... Cast it down in agriculture, mechanics, in commerce, in domestic service, and in the professions.... Our greatest danger is that in the great leap from slavery to freedom we may overlook the fact that the masses of us are to live by the productions of our hands, and fail to keep in mind that we shall prosper in proportion as we learn to dignify and glorify common labor and put brains and skill into the common occupations of life.... No race can prosper till it learns that there is as much dignity in tilling a field as in writing a poem. It is at the bottom of life we must begin, and not at the top."

Until his death in 1915, Washington was acknowledged to be the leader of black America. Graduates of Tuskegee Institute became important members of African American communities. His leadership brought positive accomplishments, but some African Americans argued that Washington had won white support by accepting an inferior position for blacks in American society.

the same thing happened the second time. I would have walked up, but I was going to the fifth story, and my long walk downtown had tired me. The third time the elevator came down the boy pointed to the sign and said, "I guess you can't read; but niggers don't ride in this elevator; we're white folks here, we are. Go to the back and you'll find an elevator for freight and niggers."...

I have been humiliated and insulted often, but I never get used to it; it is new each time, and stings and hurts more and more.

Mabel Lucas Murphy was born in 1894 in Missouri. While she was growing up, her family moved to Kansas and then Oklahoma. As an old woman, Murphy described the development of segregation.

When we moved to Kansas and Oklahoma, we had different people to encounter. They didn't have schools then like they have now. All different nationalities went to the same school. We went to the same churches. We didn't know color. I never heard my mother separate color. Never. We were the only dark people in the school and there weren't but three of us [black] children. The rest was Bohemian, Indians, Germans, Irish, Dutch. We were only three Negroes.... But after all's said and done, everybody was just lovely and we never heard no names called and no separation or nothing.

The change came in 1907, when they began to sit Negroes in one school and whites in another.... I was nine when that first separation came about. Their reason I couldn't dare tell you. It was done by the Government. My father fought it until he died. He was a teacher there, and he based his whole body and soul on reading, writing, arithmetic, and spelling.... And he thought the separation would be bad for education. But one person trying to fight it down couldn't win. You know we couldn't drink where they drank, we couldn't go to their schools, and we couldn't go to the white churches. We were all separated.

During the 1890s, lynchings became commonplace. The Reverend E. Malcolm Argyle described the situation in Arkansas in an article in the Philadelphia Christian Recorder.

There is much uneasiness and unrest all over the State among our people, owing to the fact that [black] people all over the State are being lynched upon the slightest provocation.... In the last 30 days there have been not less than eight colored persons lynched in this State. At Texarkana a few days ago, a man was burnt at the stake. In Pine Bluff a few days later two men were strung up and shot.... At Varner, George Harris was taken from jail and shot for killing a white man, for poisoning his domestic happiness. At Wilmar, a boy was induced to confess to the commission of an outrage, upon promise of his liberty, and when he had confessed, he was strung up and shot. Over in Temeoke County, a whole family consisting of husband, wife and child were shot down like dogs....

What is the outcome of all this? It is evident that the white people of the South have no further use for the Negro. He is being worse treated now, than at any other time, since the [Confederate] surrender. The white press of the South seems to be subsidized by this lawless element, the white pulpits seem to condone lynching.... The Northern press seems to care little about the condition of the Negores [in the] South. The pulpits of the North are passive. Will not some who are not in danger of their lives, speak out against the tyrannical South...speak out against these lynchings and mob violence? For God's sake, say or do something, for our condition is precarious in the extreme.

Walter White had very light skin and could easily have "passed" for white. Yet he joined the struggle for African American rights and became secretary of the NAACP in 1929. White recalled that in 1906, when he was 13, a white mob invaded his neighborhood in Atlanta. It was then that young Walter found his true identity.

Father told Mother to take my sisters, the youngest of them only six, to the rear of the house, which offered more protection from stones and bullets.... Father and I, the only males in the house, took our places at the front windows of the parlor.... In a very few minutes the vanguard of the mob, some of them bearing torches, appeared.... In a voice as quiet as though he was asking me to pass him the sugar at the breakfast table, [Father] said, "Son, don't shoot until the first man puts his foot on the lawn and then—don't you miss!"

In the flickering light the mob swayed, paused, and began to flow toward us. In that instant there opened up within me a great awareness; I knew then who I was. I was a Negro, a human being with an invisible pigmentation which marked me as a person to be hunted, hanged, abused, discriminated against, kept in poverty and ignorance, in order that those whose skin was white would have readily at hand a proof of their superiority.... No matter how low a white man fell, he could always hold fast to the smug conviction that he was superior to two-thirds of the world's population, for those two-thirds were not white....

The mob moved toward the lawn. I tried to aim my gun, wondering what it would feel like to kill a man. Suddenly there was a volley of shots. The mob hesitated, stopped. Some friends of my father's had barricaded themselves in a two-story building just below our house. It was they who had fired.... The mob broke and retreated up Houston Street.

In the quiet that followed I put my gun aside and tried to relax. But a tension different from anything I had ever known possessed me. I was gripped by the knowledge of my identity, and in the depths of my soul I was vaguely aware that I was glad of it.

W. E. B. Du Bois

Among the black leaders who felt that Booker T. Washington had sacrificed too much in return for white support was William Edward Burghardt Du Bois. Born free in Massachusetts, and a brilliant student who received a doctorate from Harvard University, Du Bois wanted to organize the best-educated African Americans to fight for full equality for all. In 1905, he and other northern blacks joined together in the Niagara Movement. Its statement of purpose read:

"We believe that [Negro] American citizens should protest emphatically and continually against the curtailment of their political rights.... We believe also in protest against the curtailment of our civil rights.... We especially complain against the denial of equal opportunities to us in economic life.... American prejudice, helped often by iniquitous laws, is making it more difficult for Negro-Americans to earn a decent living."

In 1909, Du Bois was part of a group of whites and blacks who formed the National Organization for the Advancement of Colored People (NAACP). For many years, he edited *The Crisis*, the journal of the NAACP. In later life, Du Bois sought to link the African American rights struggle with the pan-African movement, which aimed to win independence for the African nations that had been colonized by European powers. When he was 93, he went to live in the newly independent African nation of Ghana, where he died.

A bus station in Durham, North Carolina. The "colored waiting room" was a sign of the discrimination that blacks sought to escape during the Great Migration.

THE GREAT MIGRATION

Robert S. Abbott founded the Chicago Defender *in 1905. It became the most influential African American newspaper, with a circulation of over 200,000, including readers in the South as well as the North. Abbott encouraged southern blacks to migrate North. The following are some of the many letters that flowed into the editorial office asking for assistance in making the journey.*

4/20/17

Sir: I am writing you to let you know that there is 15 or 20 familys wants to come up there at once but cant come on account of money to come with...we can't phone you here we will be killed they [whites] don't want us to leave here.... if you send 20 passes [free railroad tickets that prospective employers sometimes provided for migrants] there is no doubt that every one of us will com at once. we are not doing any thing here we cant get a living out of what we do now som of these people are farmers and som are cooks barbers and black smiths but the greater part are farmers & good worker & honest poeple.... we all wants to leave here out of this hard luck place...I am a reader of the Defender and am delighted to know how times are there [in Chicago] & was glad to know if we could get some one to pass us away from here to a better land.

Dear Sirs: I am writeing to you all aksing a favor of you all. I am a girl of seventeen...I now feel I aught to go to work. And I would like very very well for you all to please forward me to a good job...I am tired down hear in this—I am afraid to say.

Dear Sir Bro.... I seen in the Defender where you was helping us a long in securing a posission as brickmason plaster cementers stone mason. I am writing to you for advice about comeing north.... We expect to do whatever you says. There is nothing here for the colored man but a hard time which these southern crackers gives us.

Northern employers sometimes opened agencies in the South to assist migrant workers who wanted northern jobs. Orlando Hubbard, who lived in Virginia, described how a labor contractor helped him to get a job in Pennsylvania.

There used to be transportation going to West Virginia, Pennsylvania, New York, and all over the country. The boss man would ask you, "Where you want to go?" You told them, and they would say, "Well, we have a special train going out of here such and such a time, you can go on that."

I made a reservation because the train was going to Homestead, Pennsylvania, to the steel mills. A bunch of fellows went. See, if you went out there and stayed ninety days you wouldn't have to pay your fare. If you decided you wanted to leave sooner they took your train fare out of your pay.

Southern employers, afraid of losing their farm laborers, sometimes took steps to prevent them from leaving. Mildred Mack Arnold was the oldest of the nine children of Edward and Minnie Mack, sharecroppers. Her family left the small town of North in South Carolina in 1924 when she was 12. Years later, Mildred Arnold recalled the difficult journey.

When my daddy got ready to leave from the South he just couldn't catch a train and leave. North was what they called in those days a "whistle-stop." It was between Florida and Columbia [South Carolina]. The special went down to Florida. If somebody had to get off at North, or if somebody was getting on, then they would stop the train. However, black people, Negroes, couldn't get on and off that train in North at that time. They couldn't do it. The white folks would stop you.

So my daddy had to leave North at night. He had to get somebody with a horse and buggy to drive him to Columbia. But the next morning the man who drove him had to be back so they wouldn't know what [he] did that night. He had to get back the next morning to go to work just like he hadn't been anyplace.

[Several months later my father] sent my mother some money and she got us all ready. It was just before the Fourth of July holiday.... There was a lady in Columbia who was like the underground railroad, like things you read about. Her place was a stopping-off place. She used to live in North and when you went to Columbia, that's where you went. Her name was Minnie Durant and she was famous for her ice cream. Everybody, if they went to Columbia, they had to get some of Miss Durant's ice cream. That's the way she made her living. When you got to Columbia, you could lay over at her place until you could get out; she would put you up.

So my mother came up to the station that Saturday morning; we were all dressed. Not too much baggage. The baggage was gone; our clothes were gone. They were taken the night before...to Miss Durant's. We had just enough like we were going up to Columbia just for the day.

We children were on the platform; I was the oldest. We didn't talk to anyone because you didn't want to let the white people know what was going on. You didn't dare. All my mother wanted to do was to get that train that morning so she could get to Columbia. So we had to be very quiet.... My aunt came to see us off and she was saying, "I'll see you Sunday when you get back. Or if I don't, I'll see you Tuesday night

Pullman Porters

Virtually all the porters on the railroad sleeping cars owned by the Pullman Company were African Americans. They played a crucial role in the Great Migration. Many personally delivered copies of the Chicago Defender *throughout the South. They also assisted the migrants whose journey to the North was usually their first train ride. Clifton L. Taulbert, who left Greenville, Mississippi, in 1963—when the migration was still going on—described the importance of the Pullman porters:*

For decades, the colored porters had served as ambassadors for colored folks who rode the train. They made sure that we understood the rules and that we safely reached our destinations. Due to their care in guiding us through the Jim Crow South, hundreds of thousands of blacks arrived unharmed.

We didn't know the porters' names. Most of them were simply called "George." Whatever the day or time, any one of the porters would answer when he heard a call for "George." The porters had names, families, and dreams, but it was a different time then. They went dutifully about their jobs answering to "George" and making life easier for thousands of railroad passengers. Though trained to serve during an era that minimized their human equality while maximizing their service skills, they were a proud cadre of professionals. Now it was my turn to be guided safely north by a kind and skillful black porter....

I eased into the comfort of the train and let the monotonous sound of the engine keep me awake as I tried to picture the transformation that would take place the moment the train crossed the Mason-Dixon line.

This group is waiting for a boat that will take them out of the South to what they hoped would be a better life.

sometimes." This kind of thing. You left the shades drawn in the house so nobody could see that the furniture and stuff were all gone....

In North you just didn't...go getting a train unless the whites were sending you someplace. They felt you belonged to them.

John H. Johnson founded the Johnson Publishing Company, whose magazines include Negro Digest, Black World, Ebony, *and* Jet. *He is today one of the nation's most successful African American business leaders. In his autobiography, Johnson remembered how his family moved from Arkansas to Chicago in 1930, when he was 12. One reason they moved was to enable John to attend high school.*

There was no Black public high school in Arkansas City. But Arkansas City, according to the *Chicago Defender* and the letters we received from the North, was not the whole world. There were Black public high schools in Chicago and other northern cities. Millions of Blacks had migrated to the North to take advantage of these facilities. And we were "free" to join them if we could save enough money to pay for the train tickets.

There were other weapons in our small arsenal. My mother's childhood friend Mamie Johnson had married "a railroad man"—Blacks who worked on the railroad were among the elite in old Black communities. The Johnsons had bought a three-flat building on Chicago's South Side. Mrs. Johnson had written several letters to my mother, suggesting that we move to Chicago and stay with her until we could get our feet on the ground. To make things even more interesting and inviting, my half sister, Beulah, was planning to move to Chicago, too....

These migrants were photographed just after their arrival in Chicago around 1912.

Like millions of other migrants, we followed the curve of the River, going from Little Rock to Memphis to St. Louis to Chicago. For most of the journey, my face was pressed so tight against the window that I could hardly breathe. I was captivated by the tall buildings, motorcars, and bright lights. I wanted to get off at Little Rock. I wanted to get off in Memphis. I had to be physically restrained in St. Louis.

When, late at night, we finally arrived at that Illinois Central Station at Twelfth Street and Michigan Avenue, four blocks from what is now my corporate headquarters, I stood transfixed on the street. I had never seen so many Black people before. I had never seen so many tall buildings and so much traffic....

We got a taxi and went to a thin, three-story building at 422 East Forty-fourth Street. My mother's friend Mamie Johnson warmly greeted us. She'd made a bedroom out of the third-floor attic. My mother slept in the bed. I slept in a rollaway bed in the same room.

Before closing my eyes on that first night, I inspected the steam heat and the inside toilet.

I was impressed.

Chicago, as the man said, was my kind of town.

Jackie Robinson, who would become the first African American player in baseball's major leagues, was born in 1919 on a plantation near Cairo, Georgia. Robinson related why his mother decided to take the family to California.

My mother is a religious woman. She had no formal education, but she possessed the wisdom to perceive that times were changing. And she was gifted with the courage to speak out in defense of her beliefs. She concluded that her four boys and one girl would never rise above their parents' status in the South. She knew that we could receive a superior education only in that part of the United States that was not corrupted by memories of the days when Negroes had been on a par with domestic animals.

To move away from the South required money, and to accumulate money was not easy. My parents labored from dawn to dark on the white man's land, then saw the products of their toil snatched from them. The plantation owner gave them a few dollars for the necessaries of life, then took the dollars away by overpricing the goods they were forced to buy at his general store. Only by the greatest thrift could my mother save enough to buy the presents which she laid at the foot of our Christmas tree. She was always in our master's debt.

I was thirteen months old when my mother finally was able to buy train tickets to California for her children and herself. That this was possible at all was due to her insistence that my father demand a larger percentage in cash for his crops.

Our destination was Pasadena, where my mother's brother, Burton, worked as a gardener.... My mother refused to accept more from him than the roof over our heads. She went into the open labor market for work as a domestic. We children seldom saw her during the day; she was gone before we were up and did not return until we were in bed. She was hands caressing us or a voice in our sleep.

The Great Migration brought many southern African American customs to the North. Among them was the outdoor barbecue. Mildred Arnold recalled:

On the Fourth of July in the South everybody would get together and have a big barbecue at the school. Everybody would bring different foods to share. My mother might bring some cakes and another woman might bring some pies. The men would have been up the whole night before the Fourth barbecuing the meat. By the time we arrived, the meat would be ready.

We brought this tradition up from the South but it was changed. Up here, we didn't have a big barbeque at the school, but families and friends would get together and have a cookout. So when I see people now with their cookouts, I smile because I know that this originated in the South.

For many of the migrants, arriving at Chicago's Illinois Central train station at 12th Street and Michigan Avenue meant that they had reached the end of their journey.

Even in the worst times of racial discrimination, some African Americans overcame the system to obtain an education and to prosper.

CHAPTER FIVE

MOVING UP

The Great Migration coincided with the growth of new national organizations aimed at improving the condition of African Americans. White and black progressives started the National Organization for the Advancement of Colored People (NAACP) in 1909. Since then, the NAACP has been in the forefront of the drive to obtain full legal rights for African Americans. It has lobbied for laws protecting the rights of African Americans, brought lawsuits to overturn Jim Crow laws, and provided legal aid for black defendants in trials that were race-related.

In 1911, the National Urban League was founded to help southern migrants adjust to conditions in northern cities. By that time, African American neighborhoods were growing on the South Side of Chicago, the East Side of Detroit, Harlem in New York City, and in Milwaukee, Pittsburgh, Cleveland, and Gary, Indiana.

The black migrants did not escape racism by moving north, however. Urban whites, alarmed by the great influx of blacks, forced them to live in restricted areas. In 1918, a leader of the Hyde Park Improvement Protective Club, in a white Chicago neighborhood, declared, "The districts which are now white must remain white. There will be no compromise." The threat was real. Chicago's South Side, the "district" for blacks, ended at 43rd Street. Beyond that line, 58 houses occupied by African Americans were fire-bombed between 1917 and 1921. In several other cities, white resentment over the influx of black families caused race riots.

Nonetheless, African Americans created a flowering of black culture in Harlem, the New York City neighborhood that became known as "the capital of Black America." From 1910 to 1930, the African American population of Harlem mushroomed to about 100,000 people. Wealthy black entertainers and doctors bought Harlem townhouses, while less prosperous residents raised rent money (especially during the depression years) by holding all-night "rent parties," where people paid admission to enter. In theaters such as the Savoy and the Apollo, black musicians from Chicago introduced the jazz music that had come up the Mississippi from New Orleans. White New Yorkers, eager to hear musicians such as Fletcher Henderson, Cab Calloway, and Thomas "Fats" Waller, began to come uptown at night. But white gangsters took advantage of Harlem's popularity, opening nightclubs such as the Cotton Club, where the audiences were white and the perfomers, black.

At least one black millionaire lived in Harlem—Madame C. J. Walker, who had earned a fortune with a chain of beauty parlors and a line of hair preparations that she had devised. After her death in 1919, her daughter A'Lelia turned her mansion, known as the Dark Tower, into a gathering place for African American writers and thinkers. Renowned authors such as Langston Hughes, Claude McKay, Countee Cullen, Arna Bontemps, Jean Toomer, and Zora Neale Hurston created poetry and novels in the era that became known as the Harlem Renaissance.

Harlem became a beacon for talented young African Americans, who dreamed of going there to taste its freedom and festivity. But only a few could make the journey. Poverty, lack of education, and racism still blasted the lives of millions of African Americans. Hemmed in by prejudice and victimized by racial violence, they survived by relying on the support of their families. Parents, grandparents, aunts, uncles, and cousins joined together to raise children and support their kin. The process of moving north often started with one family member who, after finding a job and an apartment, sent for relatives left behind.

The autobiographies of many black Americans begin with the moment in childhood when they first encountered racism. The pain worsened when the child's parents struggled to explain that the color of their skin set them apart—that it required them to ride in the back of the bus, made them the last ones served in stores, prevented them from using rest rooms or sitting in sections of theaters reserved for whites. The realization that even their own parents could do nothing about this was a crushing experience that black children never forgot.

One of the few places that African Americans could be truly free was their own church. From the time of slavery, blacks embraced Christianity, believing its message that a better life awaited them in heaven. In the late colonial period, slaves were encouraged by their owners to attend church services and often developed their own forms of devotion and hymns. As time went on, however, slave owners discouraged gatherings of slaves, even for religious purposes. After Reconstruction, many blacks remained committed to the Baptist faith in which their parents had been raised.

The first true African American church began in 1787. Richard Allen and Absalom Jones, free blacks in Philadelphia, led a prayer meeting to protest the segregated seating for blacks in a local Methodist church. By 1816, Allen had become the first bishop of the African Methodist Episcopal church, which today has more than 2 million members in 6,200 congregations.

In both the North and South, African American churches became the social centers of their communities. Here, at church suppers and socials, blacks could meet without harassment from the white world. At Sunday services, the people raised their voices in songs that derived from the African music their ancestors had brought to America.

Soon after the beginning of the Great Migration, Pentecostalism, a fundamentalist Christian movement, arose in Los Angeles. It spread rapidly to many other cities, where countless storefront churches gained adherents. The customary Pentecostal practices of dancing, call-and-response vocalization (in which the minister calls out a phrase and the congregation responds with exclamations), and the use of drums and other percussion instruments were direct reflections of African religious practices.

Allen E. Cole opened his photo studio in Cleveland in the 1920s. His photographs, including the wedding party above, were salvaged from an attic in the 1970s.

Some African Americans have also adopted a form of Islam, the religion of the rulers of the great African empires. W. D. Fard, also known as Wali Farad, is recognized as the founder of the Nation of Islam, or Black Muslims. Farad started a Temple of Islam in Detroit around 1930. Elijah Poole, the son of a Georgia tenant farmer, became a member of the temple and took the name Elijah Muhammad. He led the movement from 1934 until his death in 1975. During that time, the Nation of Islam steadily gained strength in urban black neighborhoods, where its members have established mosques, schools, businesses, and community-service centers.

Other smaller religious sects have also attracted black followers. In the 1930s, Father Divine, the son of Georgia sharecroppers, established religious communes in cities across the country in which members shared their responsibilities and incomes.

Before 1954, African American children's experiences in school did little to help them overcome the racial prejudice of society. The legally separate black school systems of the South, which received little funding, had few textbooks and almost no equipment beyond a blackboard. In the North, most black children attended local schools that were also segregated, because they were the only students in their neighborhood districts. Where African American children did go to integrated schools, they faced the same prejudice as in society at large.

Some students persevered against the odds and went to black colleges such as Fisk and Morehouse or state-sponsored colleges. By 1900, there were 34 similar institutions of higher education for blacks in the United States. Most African American lawyers, doctors, and dentists were graduates of those schools. Only a few blacks, such as W. E. B. Du Bois and the historian Carter G. Woodson (founder of Black History Month), obtained degrees from predominantly white universities before 1920.

Other schools for black students followed the model of Tuskegee Institute in Alabama. Founded by the state legislature in 1881, Tuskegee's first head was Booker T. Washington. Tuskegee trained students to become teachers (of other blacks), skilled workers, and domestic servants. The school also provided opportunities for African Americans to learn advanced methods of farming. George Washington Carver, a scientist who taught at Tuskegee, developed more than 300 uses for the peanut, donating the proceeds from his patents to the school. Carver's work proved a boon to southern farmers, both white and black.

Mary McLeod Bethune, the daughter of former slaves, was another pioneer in African American education. In 1904 she founded a school in Daytona Beach, Florida, to train young black women to become teachers. Today it is known as Bethune-Cookman College. In 1936 Bethune became the first black woman to head a federal office, the Division of Negro Affairs of the National Youth Administration.

African American workers organized labor unions as early as 1869, when delegates from 21 states formed the Negro National Labor Union. Isaac Myers, its leader, advocated cooperation with the predominantly white labor organizations of the time. Indeed, some of them, including the Knights of Labor and the Industrial Workers of the World, actively recruited black members.

However, the American Federation of Labor (AFL) barred African Americans and tried to force them out of the trades that its white members belonged to. Without a union card, African Americans were thus relegated to jobs in which they did not compete with white workers—jobs as janitors, post-office clerks, and waiters.

A picnic sponsored by the Beth Eden Baptist Church in Oakland, California, in 1905. African American churches played an important role in many black communities.

Most of the industries in which African Americans found jobs during the Great Migration were not unionized until the 1930s, which was one reason why blacks could work in them. In that decade, the United Automobile Workers organized both black and white workers to unionize the automobile industry. The Congress of Industrial Organizations (CIO), founded in 1937, adopted a nondiscriminatory policy, and by 1940 its membership included 200,000 African Americans.

The most powerful African American labor leader of the 20th century was A. Philip Randolph, founder of the Brotherhood of Sleeping Car Porters. Since the 19th century, railroads had hired African Americans as waiters in the dining cars and porters in the sleeping cars. In 1925, Randolph began to organize them, but it took 10 years for him to win recognition for the union.

Randolph's achievements did not end there. In 1941, as the United States was preparing for World War II, Randolph urged President Franklin D. Roosevelt to order a nondiscrimination policy in war-related industries and federal employment. When Roosevelt hesitated, Randolph threatened to lead a protest march of 100,000 black Americans into the nation's capital. The President relented, and when war came, blacks and whites worked side by side on the assembly lines that produced planes, tanks, and guns. However, the nondiscrimination policy did not apply to the armed forces. The one million African Americans who served in World War II fought in segregated units.

Randolph lived long enough to see that policy change, too. He was in the forefront of the postwar civil rights movement. In 1963, Randolph carried out the idea he had suggested 22 years earlier. He and Martin Luther King, Jr., led a March on Washington for Jobs and Freedom—followed by a quarter of a million people. A new era had begun for African Americans.

THE PROMISED LAND

Louis Armstrong formed his own jazz band in 1925. His long career on stage, screen, and television made him one of the most popular American entertainers of all time.

Louis Armstrong learned to play the cornet in a reform school in his native New Orleans. In 1919, when he was 19, he received a telegram from Joe "King" Oliver, whose band was playing in Chicago. Armstrong accepted Oliver's offer of a job. Many years later, Armstrong recalled his first impressions of Chicago, which were probably not very different from those of thousands of others in the Great Migration.

For the train ride up to Chicago Mayann [his mother] fixed me a big trout sandwich. And I had on my long underwear 'cause she didn't want me to catch cold. I had me a little suitcase—didn't have but a few clothes—and a little case for my cornet. I traveled in style, you know, in my way.

When I got to the station in Chicago, I couldn't see Joe Oliver anywhere. I saw a million people, but not Mister Joe, and I didn't give a damn who else was there. I'd never seen a city that big. All those tall buildings. I thought they were universities. I said, no, this is the wrong city. I was just fixing to take the next train back home—standing there in my box-back suit, padded shoulders, doublebreasted, wide-leg pants—when a redcap [train porter] Joe had left word with came up to me. He took me to the Lincoln Gardens and when I got to the door there and heard Joe and his band wailing so good, I said to myself, "No, I ain't supposed to be in *this* band. They're too good." And then Joe came out and said, "You little fool, come on in here." I said, "OK, Dad." I was home then.

A young woman at a linotype machine in a print shop. Given the chance, many like her demonstrated that their skills made them valued workers.

Maggie Comer was born in Woodland, Mississippi, the daughter of sharecroppers, and migrated to East Chicago, Indiana. She described the neighborhood in the family history Maggie's American Dream.

In August 1920, I came into Chicago on a train. Me and Carrie [her half-sister] got off the train in Chicago and we took the streetcar into East Chicago, which was about thirty miles away. I didn't stop to think too much about Chicago or East Chicago. All I wanted was freedom. I thought, "I can make it anywhere!"

East Chicago was just a few houses, and most of it was steel mill and smoke. There wasn't any finery or anything of the kind, but that wasn't what I was looking for. I was looking for a better life for myself.

Mildred Arnold came to Newark, New Jersey, from South Carolina with her family in 1924. She described her first impressions.

My father and uncle met us at the Pennsylvania Railroad Station [in Newark] and brought us up here to Newton Street, right across from the Newton Street School. We came on the trolley.... Oh Lord, to come up South Orange Avenue on that trolley car, that was something. I never rode on a trolley before. I had never even seen a trolley. I was saying to myself, "What is this? We can ride like this?"

Oh, yes. Everything was amazing.... You had a lot of gaslights in Newark. When they came on in the nighttime,...the streets lit up. We had never seen anything like that. Down South, when the sun went down there was only darkness.

William Pruitt was born in Huntsville, Alabama, and quit school in seventh grade to work as a sharecropper. In 1932, he left Alabama to work in the steel mills of Youngstown, Ohio. During the trip he spent all his money. Years later he remembered his arrival in the North.

I didn't have a penny when I got off the bus in Youngstown. I didn't even have any cigarettes, and this fellow gave me a pack of Lucky Strike cigarettes. So I couldn't get a taxi and go to my brothers' place.

So I was down there smoking the cigarettes and I walked outside the station hoping I'd see one of my brothers go by. I

A black-owned shoe shop in Junction City, Kansas, in 1915.

had told them what time I would be coming. Sure enough I saw one; he was coming to the station to see if I was there. I looked the other way and pretended I hadn't seen him.

He said, "God, Bill!"

And I looked like I hadn't seen him though I had.

He said, "How come you're standing there? Why didn't you get a cab and go on up to the house?"

I said, "I was just fixing to get one. I was kinda tired and I was just standing here looking."

Well, that's not the way it was, but I was so ashamed that I had spent all my money.

After reaching the North, many African Americans—like this man operating a drill press around 1940—found good jobs in factories.

Bernice Robinson, who came from South Carolina to New York City in the 1930s, found a society with more opportunities.

There [New York] you could go to the theater and sit anywhere you wanted to sit. You could get on the bus and sit anywhere you wanted. You had a certain amount of freedom there that [you] didn't have elsewhere. You could speak up if you didn't like something. There was subtle segregation in New York at certain exclusive restaurants. They'd just let you sit there all day and they simply wouldn't serve you. But we didn't go to them anyway since they were so expensive, except for one time when my girlfriend and I, just for fun, got these wraparound turbans and put them on our heads. [*Laughs*] We went into this segregated restaurant looking elegant. Had on big earrings and a *lot* of jewelry and stuff, and boy, we got the royal treatment! We made believe we couldn't understand exactly what was being said, and they treated us like some African dignitaries or something. [*Laughs*] And they served us....

Also in New York you could register to vote. I couldn't register in South Carolina. And blacks could run for office. In fact, my first taste of politics was through a friend of mine who was a lawyer and who had been an assemblyman for twelve years. He hired me to go over to his office after work and help mail out cards and letters to his constituents. He was originally from South Carolina, and we got to know each other because my brother lived in Sumter and we all knew some of the same people. And that was in 1944—a long time before that kind of thing happened in the south.

And on top of that, I could take my daughter to concerts and plays downtown at Carnegie Hall and Town Hall. I used to take her to hear Dean Dixon conduct. He was a black conductor who I had met personally because I did his mother's hair. We'd hear Duke Ellington—all of them. I used to try to expose her to music like that. Of course, she just liked the boogie beats then. I broke an umbrella over her head one time for slipping off and going to the Apollo Theater!

New York really prepared me to live in an integrated society. Even the work in the garment industry contributed because there you had Armenians, you had Germans, you had Jews; we *all* worked side by side on those machines.... We would go to

William O. Walker, an African American, published newspapers in Ohio's three largest cities—Cleveland, Cincinnati, and Columbus. A group of his newsboys gathers in front of the Cleveland office in 1934.

lunch together, and we would get together some evenings for movies.

The mother of John A. Williams, a writer, came from Mississippi to Syracuse, New York, in 1925. There she married and started a family. Williams fondly recalled his parents' life in the North.

My mother had come north to work for a white family. She thought the North was salvation; while the streets were not paved with gold, she sensed opportunity. In Syracuse white people did not ride down upon you at night, and they did not lynch you. But she did not know that in Syracuse the white population simply left the black population to molder in the narrow alleys along E. Washington Street, where the Negro section was.

There were few complaints of segregation or discrimination; the immigrant population the world over expects to start at the bottom and work to the top. I sometimes think that the blacks of my parents' generation were the last to believe, at least partially, that hard and honest work brought good reward. We moved about the city, within well-defined areas, most of them close to the New York Central rails that went through the heart of the city. Our moves, of course, were made hand in hand with our fortunes, and these were tied to my father's work. He was a day laborer; the label "non-skilled," bears a stigma today, but it didn't then. I live in New York City now and I don't often see men dressed in the clothes of a laborer—gray trousers, bulky, colorless sweaters, dust-lined faces, and crumpled caps or felt hats bent out of shape; nor do I smell honest sweat anymore, strong and acrid, as I used to smell it on my father and later on myself. I remember him—a

Helen Brown, who lived in New York's Harlem in the 1920s, recalled the practice of "passing," in which light-skinned African Americans posed as whites so they could obtain jobs that were not available to blacks.

Most of the people who were light-skinned were capable of going downtown and getting a job easier than others. Many people passed. McCreery's didn't hire black salespeople. I knew this girl [who worked there, even though she was black]. When her mother went down there, she couldn't speak to her. She'd pass her a note.

There was another man, who worked in Macy's. Someone [else] wanted a job, so he went to the man who was head of the department. The [head] said he didn't hire colored salesmen, but the man said, "Oh, but you do." He took him down to the shoe department and pointed out my friend who was passing. "There he is."

The manager almost died of hydrophobia. He took the salesman, and said, "Come, I want to talk to you. You didn't tell me you were colored."

He said, "You didn't ask me."

"Come and get your pay."

He left, and we don't know where he went to live.

As they moved out of the South, African American migrants brought one of the nation's greatest cultural treasures with them—jazz. Bismark Ferris (center front) started his Satisfied Orchestra Concert Band in Texas in the 1890s. After he arrived in Los Angeles a few years later, Ferris introduced his music to enthusiastic audiences.

chunky little man a bit over five feet tall, all muscle, and with the sharp face, high cheekbones, and prominent nose of an Indian—walking briskly down the street in the mornings, trailing his hand-truck behind him. He supported us on his back and muscle, his sheer strength, which was considerable.

Many African American women took domestic work, an occupation that other Americans shunned. Gertrude Dixon was born in Waycross, Georgia, around 1910. After finishing high school, she moved to Philadelphia to work as a domestic. When she was 74 she told an interviewer about her work.

A can of sardines they'd leave on the table, one egg and your bread. Now you were working all day long, washing walls and porches and all that stuff. Could you live on that?

It wasn't worthwhile, it wasn't a balanced meal. I remember one lady, every morning she'd leave a little something for me on the table. But she would tell me what they wanted for dinner. And I'd go to the frigidaire and see all this good food—chicken and steak and all those things.

So one day I cleaned that box out. What I couldn't eat I put in the garbage can, and I was gone when she got there.

The next morning, the lady didn't go to work. [*Laughs*] She said, "What happened to the chicken? What happened to the steak? Oh my God! What's the matter?"

I said, "There wasn't anybody here but me, you know that. I ate the chicken and everything else I wanted. And what I didn't want is out there, in the garbage can. I'm going to give you a lesson. Anybody come in here, you're going to feed them from now on."

She couldn't get over it. But she went around the corner to the store and told the man to let me have what I wanted to have for my lunch each day.

Race riots sometimes broke out in the cities where the African American population had grown during the Migration. One of the worst took place in East St. Louis in 1917. The following letter from a black resident to a friend described the horrors her family experienced.

Dearest Louise:

Was very glad to hear from you. Your letter was forwarded from what used to be my house.

Louise, it was awful. I hardly know where to begin telling you about it. First I will say we lost everything but what we had on and that was very little....

It started early in the afternoon. We kept receiving calls over the phone to pack our trunks and leave, because it was going to be awful. We did not heed the calls, but sent grandma & the baby on to St. Louis & said we would "stick" no matter what happened. At first, when the fire started, we stood on Broadway & watched it. As they neared our house we went in & went to the basement. It was too late to run then. They shot

& yelled something awful, finally they reached our house. At first, they did not bother us (we watched from the basement window), they remarked that "white people live in that house, this is not a nigger house." Later, someone must have tipped them that it was a "nigger" house, because, after leaving us for about 20 min[utes] they returned & yelling like mad "kill the 'niggers,' burn that house."

It seemed the whole house was falling in on us. Then some one said, they must not be there, if they are they are certainly dead. Then some one shouted "they are in the basement. Surround them and burn it down." Then they ran down our steps. Only prayer saved us, we were under tubs & any thing we could find praying & keeping as quiet as possible, because if they had seen one face, we would have been shot or burned to death.... Sister tipped the door to see if the house was on fire. She saw the reflection of a soldier on the front door—pulled it open quickly & called for help. All of us ran out then & was taken to the city hall for the night—(jist as we were)....

On Tuesday evening...our house was burned with two soldiers on guard.... We were told that [the crowd] looted the house before burning it.

Outside the South, African Americans took advantage of the right to vote. Here, members of the League of Women Voters distribute information on how to register.

Migrants wrote letters back home to describe their successes. The following letter was to a family in Mississippi.

I was promoted on the first of the month I was made first assistant to the head carpenter...and was raised to $95 a month.... What's the news generally around H'burg [Hattiesburg]? I should have been here 20 years ago. I just begin to feel like a man. It's a great deal of pleasure in knowing that you have got some privilege. My children are going to the same school with the whites and I dont have to umble [humble myself] to no one. I have registered—Will vote the next election and there isnt any "yes sir" and "no sir"—its all yes and no and Sam and Bill.

Liney Jones Rozier and her family operated this store in Los Angeles in 1906.

A couple dances the jitterbug, a popular dance of the 1940s, at a community center in Chicago.

NEIGHBORHOODS

The poet Langston Hughes described his impressions of Chicago, where he visited his mother during his summer vacation.

I went to join my mother in Chicago. Dad and my mother were separated again, and she was working as cook for a lady who owned a millinery shop in the Loop, a very fashionable shop where society leaders came by appointment and hats were designed to order. I became a delivery boy for that shop. It was a terrifically hot summer, and we lived on the crowded Chicago South Side in a house next to the elevated. The thunder of the trains kept us awake at night. We could afford only one small room for my mother, my little brother, and me.

South State Street was in its glory then, a teeming Negro street with crowded theaters, restaurants, and cabarets. And excitement from noon to noon. Midnight was like day. The street was full of workers and gamblers, prostitutes and pimps, church folks and sinners. The tenements on either side were very congested. For neither love nor money could you find a decent place to live. Profiteers, thugs, and gangsters were coming into their own. The first Sunday I was in town, I went out walking alone to see what the city looked like. I wandered too far outside the Negro district, over beyond Wentworth, and was set upon and beaten by a group of white boys, who said they didn't allow niggers in that neighborhood. I came home with both eyes blacked and a swollen jaw. That was the summer before the Chicago riots [of 1919].

Though Jackie Robinson's family had moved from Georgia to Pasadena, California, he still experienced racial prejudice.

Pasadena regarded us as intruders. My brothers and I were in many a fight that started with a racial slur on the very street we lived on. We saw movies from segregated balconies, swam in the municipal pool only on Tuesdays, and were permitted in the YMCA on only one night a week. Restaurant doors were slammed in our faces. In certain respects Pasadenans were less understanding than Southerners and even more openly hostile.

Such was my early childhood. I learned my lesson then. My mother taught us to respect ourselves and to demand respect from others.

That's why I refused to back down in later life and won't back down now....

When I was about eight I discovered that in one sector of life in Southern California I was free to compete with whites on equal terms—in sports. I played soccer on my fourth-grade

Restaurants in black neighborhoods provided soul food, which originated in the rural communities of the South. Daddy Grant's Oldtime Pit Barbecue was a popular eating place in Los Angeles in the 1940s.

team against sixth-graders who were two or three years older than me. Soon I was competing in other sports against opponents of every size, shape...and color.... The more I played the better I became—in softball, hard ball, football, basketball, tennis, table tennis, any kind of game with a ball....

In primary and high school white boys treated me as an equal. I was their star, the leading scorer in basketball, quarterback in football, a .400 hitter in baseball, the best broad-jumper on the track team. They made me a professional at eight by bringing me lunches as a bribe to play on their teams.

Daisy Lee Gatson Bates, a civil rights leader who played an important role in the integration of Little Rock's Central High School in 1957, was born in 1920. In her autobiography, she described her hometown of Huttig in southern Arkansas.

I was born Daisy Lee Gatson in the little sawmill town of Huttig, in southern Arkansas. The owners of the mill ruled the town. Huttig might have been called a sawmill plantation, for everyone worked for the mill, lived in houses owned by the mill, and traded at the general store run by the mill.

The hard, red clay streets of the town were mostly unnamed. Main Street, the widest and longest street in town, and the muddiest after a rain, was the site of our business square. It consisted of four one-story buildings which housed a commissary and meat market, a post office, an ice cream parlor, and a movie house. Main Street also divided "White Town" from "Negra Town." However, the physical appearance of the two areas provided a more definite means of distinction.

The Negro citizens of Huttig were housed in rarely painted, drab red "shotgun" houses, so named because one could stand in the front yard and look straight through the front and back doors into the back yard. The Negro community was also provided with two church buildings of the same drab red exterior, although kept spotless inside by the Sisters of the church, and a two-room schoolhouse equipped with a potbellied stove that never quite succeeded in keeping it warm.

On the other side of Main Street were white bungalows, white steepled churches and a white spacious school with a big lawn. Although the relations between Negro and white were cordial, the tone of the community, as indicated by outward appearances, was of the "Old South" tradition.

Zora Neale Hurston, born in 1901, grew up to be a writer and anthropologist. Her family lived in Eatonsville, Florida, the first black town formally incorporated in the United States. She remembered her childhood home.

We lived on a big piece of ground with two big chinaberry trees shading the front gate and Cape jasmine bushes with hundreds of blooms on either side of the walks. I loved the fleshy, white, fragrant blooms as a child but did not make too much of them. They were too common in my neighborhood. When I got to New York and found that the

Richard Wright, born in 1908 in Mississippi, moved to Chicago when he was 19 and found work as a postal clerk. During the 1930s he went to New York, where he began his writing career. His novel Native Son, *published in 1940, has become a classic. The following year he and Edwin Rosskam wrote* 12 Million Black Voices, *a history of the Great Migration. They described the little apartments where the migrants in Chicago lived:*

The bosses of the Buildings take these old houses and convert them into 'kitchenettes,' and then rent them to us at rates so high that they make fabulous fortunes before the houses are too old for habitation. What they do is this: they take, say, a seven-room apartment, which rents for $50 a month to whites, and cut it up into seven small apartments, of one room each; they install one small gas stove and one small sink in each room. The Bosses of the Buildings rent these kitchenettes to us at the rate of, say, $6 a week. Hence, the same apartment for which white people—who can get jobs anywhere and who receive higher wages than we—pay $50 a month is rented to us for $42 a week! And because there are not enough houses for us to live in, because we have been used to sleeping several in a room on the plantations in the South, we rent these kitchenettes and are glad to get them. These kitchenettes are our havens from the plantations in the South. We have fled the wrath of Queen Cotton and we are tired.

Sometimes five or six of us live in a one-room kitchenette, a place where simple folk such as we should never be held captive. A war sets up in our emotions: one part of our feelings tells us that it is good to be in the city, that we have a chance at life here, that we need but turn a corner to become a stranger, that we no longer need bow and dodge at the sight of the Lords of the Land. Another part of our feelings tells us that, in terms of worry and strain, the cost of living in the kitchenette is too high, that the city heaps too much responsibility upon us and gives too little security in return.

More than 200,000 African Americans served in the armed forces of the United States during World War I. Yet they faced the same discrimination they experienced as civilians; African American soldiers fought in segregated units, and the few accepted into the navy were cooks or stevedores.

On July 28, 1917, shortly after the United States entered the war, 15,000 African Americans marched down New York's Fifth Avenue to protest the discrimination and lynchings the people of their race faced throughout the country. The marchers distributed a pamphlet explaining their motives:

We march because we want to make impossible a repetition of Waco, Memphis, and East St. Louis [the sites of anti-black race riots] by arousing the conscience of the country, and to bring the murderers of our brothers, sisters and innocent children to justice.

We march because we deem it a crime to be silent in the face of such barbaric acts.

We march because we are thoroughly opposed to Jim Crow cars, segregation, discrimination, disfranchisement, lynching, and the host of evils that are forced on us....

We march because we want our children to live in a better land and enjoy fairer conditions than have fallen to our lot.

Women volunteers at a United Service Organization (USO) club entertain black soldiers about to leave for France during World War I. African Americans still fought in segregated units, as they would continue to do until the 1950s. But for the first time, black officers led the enlisted men into combat.

people called them gardenias, and that the flowers cost a dollar each, I was impressed. The home folks laughed when I went back down there and told them. Some of the folks did not want to believe me. A dollar for a Cape jasmine bloom! Folks up north there must be crazy.

There was plenty of orange, grapefruit, tangerine, guavas and other fruits in our yard. We had a five-acre garden with things to eat growing in it, and so we were never hungry. We had chicken on the table often; home-cured meat, and all the eggs we wanted. It was a common thing for us smaller children to fill the tea-kettle full of eggs and boil them, and lay around in the yard and eat them until we were full. Any left-over boiled eggs could always be used for missiles. There was plenty of fish in the lakes around the town, and so we had all that we wanted. But beef stew was something rare. We were all very happy whenever Papa went to Orlando and brought back something delicious like stew-beef. Chicken and fish were too common with us. In the same way, we treasured an apple. We had oranges, tangerines and grapefruit to use as hand-grenades on the neighbors' children. But apples were something rare. They came from way up north.

Elton Fax, a writer, recalled what Harlem meant to him as a boy growing up in Maryland, and the joy he felt when he finally arrived there.

When I was growing up in Baltimore in the '20s, Harlem epitomized a kind of freedom that we did not know: "Once I get to Harlem, I won't need to worry about anything. Nobody's gonna bother me in Harlem." We found out later, of course, that this was not as real as we imagined it to be. Harlem was more glamorous than the black neighborhoods of Baltimore, or of Norfolk, Virginia, or of Jackson, Mississippi, and it was known all over. Everybody wanted to see this tremendous community....

I wanted to see Harlem, of course I did. I saw it first in the '30s, and it was fabulous. Let me tell you one of the things that meant so much to simple people. There was this big black giant of a policeman, Lacy, at 125th Street and Seventh Avenue, directing traffic, making white folks stop and go at his bidding. Where I came from, there was no black authority or authoritative figure who commanded respect. To see him pull over a car full of white people and *saunter*. He was quite dramatic, and he'd say, "As big and black as I am, you mean to tell me you can't see my hand in the air?" This symbolized, for us, something that we had not seen before, and while it was superficial, it was impressive.

Lofton Mitchell, a playwright and essayist, described growing up in Harlem in the 1930s.

The small town of black Harlem, though surrounded by hostility, was crowded with togetherness, love, human warmth, and neighborliness. Southern Negroes fled from physical lynchings and West Indians from economic lynchings. They met in the land north of 110th Street and they brought with them their speech patterns, folkways, mores, and their dogged determination. They brought, too, their religiosity and their gregariousness and they created here a distinct nation that was much like a small town. Readily welcomed were newly arrived relatives and strangers, and these were maintained until they found jobs and homes. In this climate everyone knew everyone else. A youngster's misbehavior in any house earned him a beating there plus one when he got home. In this climate the cooking of chitterlings [fried pig intestines] brought a curious neighbor to the door: "Mrs. Mitchell, you cooking chitterlings? I thought you might need a little cornbread to go with 'em." A moment later a West Indian neighbor appeared with rice and beans. Another neighbor followed with some beer to wash down the meal. What started as a family supper developed into a building party.

This climate created in Harlem a human being with distinct characteristics. The child of Harlem had the will to survive, to "make it." He was taught, "If you're going to be a bum, be the best bum." He was taught, too, a burning distrust for whites, to strike them before they struck, that to "turn the other cheek" was theologically correct, but physically incorrect in dealing with white folks. He learned to hold out his right hand, but to clench his left hand if the "flesh wasn't pinched in a decent handshake." This Harlem child learned to laugh in the face of adversity, to cry in the midst of plentifulness, to fight quickly and reconcile easily....

He was poor, but proud. He hid his impoverishment with clothes, pseudo-good living, or sheer laughter. Though he complained of being broke, he never admitted his family was poor.... This Harlem child had to have something—a car, a sharp wit, sharp clothing, a ready laugh, a loud voice, a beautiful woman. He had the burning need to belong to something, own something, and let the world know: "This is me! I'm somebody!"

The poet Langston Hughes was one of the most important figures of the Harlem Renaissance. He described its spirit and hardships:

When I came back to New York in 1925 the negro Renaissance was in full swing. Countee Cullen was publishing his early poems. Zora Neale Hurston, Rudolph Fisher, Jean Toomer, and Wallace Thurman were writing. Louis Armstrong was playing, Cora Le Redd was dancing, and the Savoy Ballroom was open with a specially built floor that rocked as the dancers swayed. Alain Locke was putting together *The New Negro* [an anthology of black literature]. Art took heart from Harlem creativity. Jazz filled the night air—but not everywhere—and people came from all around after dark to look upon our city within a city, Black Harlem. Had I not to earn a living, I might have thought it even more wonderful. But I could not eat the poems I wrote.

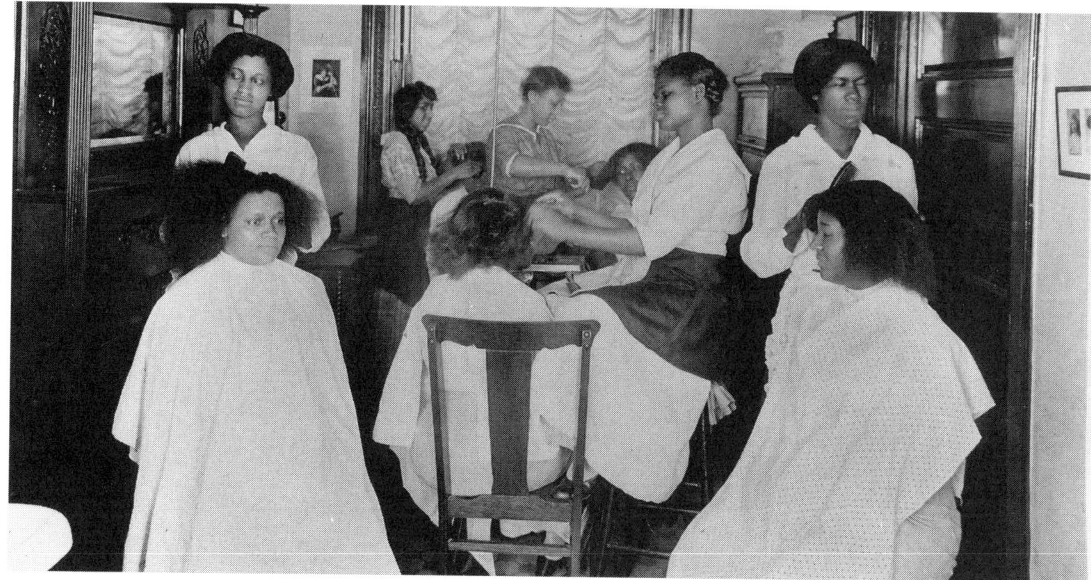

A black-owned beauty parlor in New York City's Harlem in the 1920s. Madame C. J. Walker, the first African American millionaire, made her fortune by establishing a chain of such parlors in several cities.

FAMILY

The photographer and movie director Gordon Parks recalled his childhood in the little town of Fort Scott, Kansas.

My mother, Sarah Parks, saw to it that her children ate regularly, and my father, Jackson Parks, worked the field around our small clapboard house to make that possible. He grew corn, beets, turnips, potatoes, collard greens and tomatoes. A few ducks and chickens supplied eggs, and my father always managed to have a hog to slaughter for the smokehouse, which was small and crude but served the purpose....

I will always consider my parents to be my just heroes. My fourteen siblings would, I am sure, have agreed with me. They made life more tolerable for all of us with their compassion and generosity. Yet neither of them would have thought what they did to earn our infinite respect was in any way extraordinary. Sarah Parks would have defied God Himself if what He willed her to do would harm another human being. She would have exonerated her disobedience by concluding that God had, for an instant, lost sight of his own teachings. Certainly, the devil found her a terrible enemy. Neither would she allow racism to drag her reasoning into the throes of its darkness. Without considering the consequences, she once took a homeless white child into our house to feed and clothe until a distant relative came to his rescue. That our black friends and neighbors disapproved of her actions made no difference.... The sick or disabled, no matter what their color, found her at their bedside. And although we were dismally poor, she always scraped up a basket of food to take along. She was a thrifty woman as well. One of my brothers sent me an old bank book of my mother's that he had kept for many years. Between 1912 and 1925 she had managed to save ten dollars and eighty-two cents. I looked at it in the abundant light flowing through the windows of my apartment and for several moments I stood speechless. How, I wondered, had she managed all that bountiful food for our table, those sizzling pans of baked beans, sweet potatoes, apple pies and cobblers.

Joseph P. George, his wife, Armand, and their children in Los Angeles in the early 20th century.

Birdie Lee May grew up in Arkansas in the early years of the 20th century. Because her parents worked long hours, she had to assume some responsibility for her brothers and sisters.

I was born in Menifee in Conway County, Arkansas, December 6, 1904.... Father was a sharecropper. I stayed home there and took care of the next children. I wanted to go to the fields because I thought it would be more fun than staying in the log cabin all day. They grew cotton, corn, sorghum—you know what this is, sticky like molasses. I more or less kept the children clean and out of things they shouldn't do. There were three or four under me and that was enough! I had to have the meals fixed for when my parents came home. I did this until I was eleven or twelve. One year we moved so many times it was sort of pitiful. Lived in a little log cabin for a while, then we'd see another house that was better and move to it. It was pathetic. When I think of it now, the houses we have lived in. I have known some not to have a kitchen floor. Just ground. Didn't live there too long. Father built a little house after that. At the time I thought it was wonderful, but no one now would appreciate it. Mother was happy about it.

I have known times when Mother and Father were living together on a farm and got up at four in the morning to go to the field. Sometimes I could go with them and sometimes we had to walk five or six miles—or it seemed that long! In that farm it was awful chilly. Mother would carry vegetables to the field to cook and we'd eat it there. Eat by a shade tree by the road.

I liked to keep up with my brother. We would climb trees even though it wasn't too ladylike. We never had a radio while I was a child. We would sit around and look through thick catalogues, like Montgomery and Ward's, and look at pictures and wish for this and that.

This unidentified family in Washington, D.C., around 1900 was obviously prosperous. The parents may have been graduates of the capital city's Howard University, which trained many blacks to be doctors, surgeons, professors, dentists, lawyers, engineers, and architects.

Bernice Robinson was born in Charleston, South Carolina, in 1914. She described her family's close ties.

My childhood was a happy childhood. We weren't rich. My daddy would refuse to let us say that we were poor, but he didn't make very much money. He was a bricklayer and tile setter and a plasterer—that sort of thing. My mother just took care of the home and children and did sewing, and she was the stabilizing factor, I guess, in the home. She was always quiet and soft-spoken.

My father was a very strong man, and most people were afraid of him because he always looked so stern. He was of Indian descent—Cherokee—and he very rarely smiled. He just went around looking mad all the time....

Even though my father looked stern, he had a big heart and he was concerned about people at all times. He always said you never had so little that you didn't have enough to share. And that has stayed with me over the years.

One of the first examples of that kind of thing that I remember was when I was in the third grade. We had a contest that day writing all the capitals in the United States, and we had to write them with ink and the straight pen you used then, and my paper was correct. I had no mistakes at all. There was another girl in the class whose parents had told her that she would have to win that contest. I was never pressured like that by my parents. The prize was just a pencil box. It was no elaborate award, but it was just the idea of giving us some incentive. And this girl had a small *o* for Ohio, and, of course, that was one point against her. I had a perfect score.

Well, she got together her little group to take this pencil box from me.... I was scared because I had never been in anything like that before. But there was this girl—big for her age—who followed me all the way home. She said to them, "You are not going to touch a strand of hair on her head." Her name was Jenny, and we used to call her the bully, and I couldn't understand why she was protecting me. But she walked me all the way home, which was several blocks past her house. When we got just about half a block from our house, she said, "You don't know why I stood up for you, do you?" And then she said..., "My mother is in your father's class at church, and we never knew what Christmas was like because we had never had any money to have anything, but your father brought us some toys and fruits for Christmas, and that was the best Christmas I ever had in my life. Nobody was going to touch you." And I have never forgotten that.... My father was a man like that.

Burdie Thompson (left), her husband, Tracy, and their daughter Betty in Pocatello, Idaho, in the 1930s.

The John W. Fleming family of Cincinnati, Ohio, gathers for a reunion. Among the family members present were a minister, lawyers, and college teachers.

In the segregated South, the black family had to try to shield its members from the cruelties of the society. Daisy Bates recalled the effect her first experience with racism had on her family.

As I grew up in this town [Huttig, Arkansas], I knew I was a Negro, but I did not really understand what that meant until I was seven years old. My parents, as do most Negro parents, protected me as long as possible from the inevitable insult and humiliation that is, in the South, a part of being "colored."

I was a proud and happy child—all hair and legs, my cousin Early B. used to say—and an only child, although not blessed with the privilege of having my way. One afternoon, shortly after my seventh birthday, my mother called me in from play.

"I'm not feeling well," she said. "You'll have to go to the market to get the meat for dinner."

I was thrilled with such an important errand. I put on one of my prettiest dresses and my mother brushed my hair. She gave me a dollar and instructions to get a pound of center-cut pork chops. I skipped happily all the way to the market.

When I entered the market, there were several white adults waiting to be served. When the butcher had finished with them, I gave him my order. More white adults entered. The butcher turned from me and took their orders. I was a little annoyed but felt since they were grownups it was all right. While he was waiting on the adults, a little white girl came in and we talked while we waited.

The butcher finished with the adults, looked down at us and asked, "What do you want, little girl?" I smiled and said, "I told you before, a pound of center-cut pork chops." He snarled, "I'm not talking to you," and again asked the white girl what she wanted. She also wanted a pound of center-cut pork chops.

"Please may I have my meat?" I said, as the little girl left.

The butcher took my dollar from the counter, reached into the showcase, got a handfull of fat chops and wrapped them up. Thrusting the package at me, he said, "Niggers have to wait 'til I wait on the white people. Now take your meat and get out of here!" I ran all the way home crying.

When I reached the house, my mother asked what had happened. I started pulling her toward the door, telling her what the butcher had said. I opened the meat and showed it to her. "It's fat, Mother. Let's take it back."

"Oh, Lord, I knew I shouldn't have sent her. Stop crying now, the meat isn't so bad."

"But it is. Why can't I take it back?"

"Go out on the porch and wait for Daddy." As she turned from me, her eyes were filling with tears.

When I saw Daddy approaching, I ran to him, crying. He lifted me in her arms and smiled. "Now, what's wrong?" When I told him, his smile faded.

"And if we don't hurry, the market will be closed," I finished.

"We'll talk about it after dinner, sweetheart." I could feel his muscles tighten as he carried me into the house.

Dinner was distressingly silent. Afterward my parents went into the bedroom and talked. My mother came out and told me my father wanted to see me. I ran into the bedroom. Daddy sat there, looking at me for a long time. Several times he tried to speak, but the words just wouldn't come. I stood there, looking at him and wondering why he was acting so strangely. Finally he stood up and the words began tumbling from him. Much of what he said I did not understand. To my seven-year-old mind he explained as best he could that a Negro had no rights that a white man respected.

He dropped to his knees in front of me, placed his hands on my shoulders, and began shaking me and shouting.

"Can't you understand what I've been saying?" he demanded. "There's nothing I can do! If I went down to the market I would only cause trouble for my family."

As I looked at my daddy sitting by me with tears in his eyes, I blurted out innocently, "Daddy, are you afraid?"

He sprang to his feet in an anger I had never seen before. "Hell, no! I'm not afraid for myself, I'm not afraid to die. I could go down to that market and tear him limb from limb with my bare hands, but I was afraid for you and your mother."

That night when I knelt to pray, instead of my usual prayers, I found myself praying that the butcher would die. After that night we never mentioned him again.

Gordon Parks took this picture of a family in Washington, D.C., in 1942. Parks' striking photos of urban black life won him a job with Life *magazine in 1950. Later, he became the first African American to direct full-length films for a major Hollywood studio.*

Gonzella Sullivan, a West Virginia miner, with his son in 1946.

Marian Wright Edelman

Marian Wright Edelman has been one of the most forceful crusaders for the welfare of children in America. Born in South Carolina in 1939, she was deeply influenced by her father, a Baptist minister. He raised his children to follow Booker T. Washington's philosophy of self-help. "Working for the community was as much a part of our existence as eating and sleeping and church," she recalled.

After graduating first in her class at Spelman College in 1960, Edelman earned a degree at Yale Law School. She opened a law office in Jackson, Mississippi, where she was active in the civil rights movement. Threatened and even thrown into jail, she persevered and became the first black woman to pass the Mississippi bar exam. In 1968, she married Peter Edelman, an aide to Senator Robert Kennedy. They have three sons.

In 1973 Marian Wright Edelman founded the Children's Defense Fund, a nonprofit child advocacy organization. The fund has effectively advocated legislation and social programs on such issues as adolescent pregnancy prevention, child health education, child care, and youth employment.

Edelman's work and talents brought her invitations to serve on the boards of Yale University, the NAACP Legal Defense and Education Fund, the United States Committee for UNICEF, and many other organizations. She won a MacArthur Foundation Prize Fellowship in 1985, an award given to those who have made notable contributions to American life. She has also been an influential adviser to the Clinton administration.

James Baldwin, in his novels and essays, continually searched his own soul to reveal the anguish of being an African American. Baldwin's father had been a minister in Harlem. In an essay about his father's funeral, which took place amidst a race riot, Baldwin reflected on their relationship.

I had not known my father very well. We had got on badly, partly because we shared, in our different fashions, the vice of stubborn pride. When he was dead I realized that I had hardly ever spoken to him. When he had been dead a long time I began to wish I had....

He was, I think, very handsome. I gather this from photographs and from my own memories of him, dressed in his Sunday best and on his way to preach a sermon somewhere, when I was little.... But he looked to me, as I grew older, like pictures I had seen of African tribal chieftains: he really should have been naked, with war-paint on and barbaric mementos, standing among spears. He...was certainly the most bitter man I have ever met; yet it must be said that there was something else in him, buried in him, which lent him his tremendous power and, even, a rather crushing charm. It had something to do with his blackness, I think—he was very black—with his blackness and his beauty, and with the fact that he knew that he was black but did not know that he was beautiful. He claimed to be proud of his blackness but it had also been the cause of much humiliation and it had fixed bleak boundaries to his life. He was not a young man when we were growing up and he had already suffered many kinds of ruin; in his outrageously demanding and protective way he loved his children, who were black like him and menaced, like him; and all these things sometimes showed in his face when he tried, never to my knowledge with any success, to establish contact with any of us. When he took one of his children on his knee to play, the child always became fretful and began to cry; when he tried to help one of us with our homework the absolutely unabating tension which emanated from him caused our minds and our tongues to become paralyzed, so that he, scarcely knowing why, flew into a rage and the child, not knowing why, was punished.... He had lived and died in an intolerable bitterness of spirit and it frightened me, as we drove him to the graveyard through those unquiet, ruined streets, to see how powerful and overflowing this bitterness could be and to realize that this bitterness now was mine.

Clifton L. Taulbert grew up in the segregated society of the Mississipi Delta region in the 1950s. He wrote about the closeness and warmth of his extended family, particularly his great-grandfather, called "Poppa," who raised him.

When I think of Poppa today, I am reminded of a colored southern Buddha. He was robust, very imposing and his head was as clean and shiny as that of an ancient Chinese god. Being a well-known and respected Baptist preacher, he was looked to for his wisdom and in many instances served as a go-between for the coloreds when problems arose involving whites....

Poppa was more than my best friend; he was also the essence of Christmas. For colored children in Glen Allan, Christmas was the only time of year we got fresh fruit and toys. We each got one toy at Christmas, and today it seems amazing how those toys would last forever—or at least 364 days until Christmas came again. December 24 was a big night for us, and we didn't care whether Santa Claus was white, green or yellow, just as long as he came down our chimney. Poppa always made sure that something special happened in our lives at Christmas. We'd go around to his house early on Christmas morning and he'd have eggnog and bowls of fresh apples and oranges and pecans and walnuts—all the things we rarely saw during the rest of the year. On Christmas it seemed we stepped into a fantasy land of new toys and good food, and Poppa was at its center.

The writer John A. Williams described his family life in Syracuse during the depression.

I had no clash of ill will with my father, for he left the running of the house and the disciplining to my mother. He was a soft-hearted man; not a coward, just soft. The world was running right up his back and with all those mouths to feed, and he discovered he could not fill them. For a man who thought of himself as a man, he realized that he was failing that most basic of manly tasks, to provide for his family. He was not privy to all the sociological and psychological information about why he could not; he might have blamed it on being black, and that was indeed correct. But there was also the Depression and he could not handle both. And so he left because, being a man who for some reason was not able to perform as one, he was too filled with shame to remain with us.

Long before he left, however, my father and I had good times. He was a sports fan from his chitlins out; they don't put them together that way anymore. I don't know how many Sundays we spent walking or riding the streetcar to some far section of the city to watch a baseball or football game. Even today my father will not tolerate being disturbed on Saturday or Sunday afternoons when the football games are on television.

It took me some time to realize it, but both my parents were strong people; they had no choice, they had to be. But then, most blacks have had to be strong and were. Oh, they worked the most menial jobs, performed the toughest labor, but I remember laughter and parties and singing and dancing; remember picnics and loud voices; suits and dresses carrying the odor of just coming out of the cleaners that afternoon. All was not totally grim; life bubbled, or forever sought to, beneath the hard grind of everyday life.

In 1981, Marian Wright Edelman published An Advocacy Agenda for Black Families and Children. *In it, she wrote:*

There are no miracles on the horizon to make the dream of equal opportunity a reality for Black children.... We must each take the responsibility for lifting ourselves and bringing along others. Everything we have earned as a people—even that which is our own by right—has come out of long struggle....

Sojourner Truth...pointed the way for us.... Once a heckler told Sojourner that he cared no more for her talk "than for a fleabite." "Maybe not," was her answer, "but the Lord willing, I'll keep you scratching."

Her retort should be ours today and tomorrow to a nation that keeps turning its back on our children.... Black parents and leaders have got to "make them scratch" all over the nation. Every single person can be a flea and can bite.... Enough fleas for children can make even the biggest dogs mighty uncomfortable. If they flick some of us off and others of us keep coming back, we will begin to get our children's needs heard and attended to.

A California family in the early 20th century. By 1856, African Americans in San Francisco had established their own churches, schools, a library, and a newspaper. Twenty years later, the state supreme court affirmed their right to attend integrated public schools.

RELIGION

Maggie Comer moved to East Chicago, Indiana, in 1920. She later told her daughter about the importance of the church in the community.

Life was mostly church. You went to church on Sunday from morning until eleven o'clock Sunday night. You'd go for dinner and back to church to night service. And then during the week you had about three nights that you went to choir rehearsal and prayer service. Your recreation was really church and church socials.

I didn't have a lot of close friends.... But the girls I knew were from around the church.... To go around with the preacher's daughter was something. She was the leader of the group of girls, the one that we all looked up to. Now, you had to be a very nice girl to be able to go in her group. I didn't get to spend a lot of time with them because I usually had to work on Sunday when I worked steady. But even then I was sometimes off at twelve and I could go to three o'clock service, or I was off at five and I could go to seven o'clock service.

That's how I met your dad, at church with this group of girls. He was looking for the nice kind of girl that was in this preacher's daughter's group.

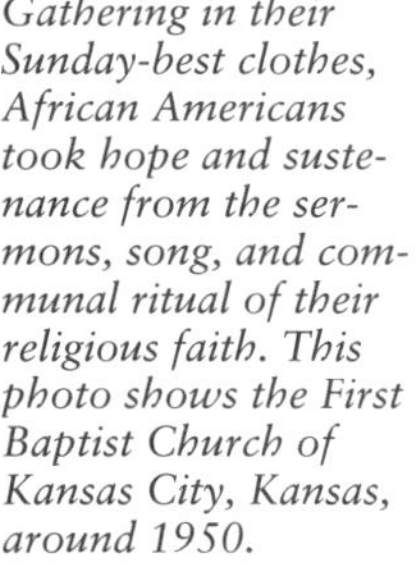
Gathering in their Sunday-best clothes, African Americans took hope and sustenance from the sermons, song, and communal ritual of their religious faith. This photo shows the First Baptist Church of Kansas City, Kansas, around 1950.

While in prison, Malcolm Little (later known as Malcolm X) received letters from his brother Wilfrid, who had become a member of the Nation of Islam, or the Black Muslims. After his release, Malcolm lived with his brother's family in Detroit. In his autobiography, he described the appeal of the Black Muslims.

I never had seen any Christian-believing Negroes conduct themselves like the Muslims.... The men were quietly, tastefully dressed. The women wore ankle-length gowns, no makeup, and scarves covered their heads. The neat children were mannerly not only to adults but to other children as well.

I had never dreamed of anything like that atmosphere among black people who had learned to be proud that they were black, who had learned to love other black people instead of being jealous and suspicious. I thrilled to how we Muslim men used both hands to grasp a black brother's both hands, voicing and smiling our happiness to meet him again. The Muslim sisters, both married and single, were given an honor and respect that I'd never seen black men give to their women, and it felt wonderful to me. The salutations which we all exchanged were warm, filled with mutual respect and dignity: "Brother"..."Sister"..."Ma'am"..."Sir." Even children speaking to other children used these terms. Beautiful!

Clifton L. Taulbert grew up in the segregated Mississippi of the 1950s. In his memoir When We Were Colored, *he described the importance of the church in his small town of Glen Allan.*

It was closer to our hearts than our homes—the colored church. It was more than an institution, it was the very heartbeat of our lives. Our church was all our own, beyond the influence of whites, with its own societal structure.

Even when colored people moved north, they took with them their church structure. The Baptist church, of which my family was a part, had (and has) a big network under the auspices of the National Baptist Convention. A small colored Baptist church in Glen Allan, Mississippi, and a large colored Baptist church in Saint Louis, Missouri, had the same moderator, used the same Sunday-school books, and went to the same conferences. And whether north or south, large or small, the colored church was a totally black experience....

As early as I can remember, I spent my Sundays, both night and day, attending church.... No matter the hard workweek, we all looked forward to Sunday when we would dress in our best and meet our friends....

Today, field hands were deacons, and maids were ushers, mothers of the church, or trustees. The church transformed the ordinary into an institution of social and economic significance. A hard week of field work forgotten, the maid's aprons laid to rest, and the tractors in the shed, these colored men and women had entered a world that was all their own. Rough hands softened with Royal Crown grease were positioned to praise. As a young boy, I sat quietly in my seat and waited for the services to start. The church was designed for us children to

The A.M.E. Church

The African Methodist Episcopal (A.M.E.) church, the first organized African American church, was formed in Philadelphia in 1816. Its first bishop was Richard Allen. Allen recalled that the church began after an incident in a regular Methodist church in 1787:

A number of us [free blacks] usually attended St. George's church in Fourth street; and when the colored people began to get numerous in attending the church, they moved us from the seats we usually sat on, and placed us around the wall. [One Sunday, the sexton of the church accidentally directed the Blacks to sit in the regular section.] We took those seats. Meeting had begun, and...just as we got to the seats, the elder said, "Let us pray." We had not been long upon our knees before I heard considerable scuffling and low talking. I raised my head up and saw one of the trustees, H—M—, having hold of...Absalom Jones, pulling him up off of his knees, and saying, "You must get up—you must not kneel here." Mr. Jones replied, "Wait until prayer is over." Mr. H—M— said, "No, you must get up now, or I will call for aid and force you away." Mr. Jones said, "Wait until prayer is over, and I will get up and trouble you no more." [H—M—, with the help of another trustee, then began to pull another black worshipper from his place.] By this time prayer was over, and we all went out of the church in a body....

We were filled with fresh vigor to get a house erected to worship God in.... We then hired a store-room, and held worship by ourselves. Here we were pursued with threats...but we believed the Lord would be our friend. We got subscription papers out to raise money to build the house of the Lord. [Two white Philadelphians, Benjamin Rush and Robert Ralston] pitied our situation, and subscribed largely toward the church, and were very friendly towards us, and advised us how to go on.... They were the first two gentlemen who espoused the cause of the oppressed, and aided us in building the house of the Lord for the poor Africans to worship in. Here was the beginning and rise of the first African church in America.

Over 11 million African Americans, more than one-third of the total black population, belong to one of the branches of the Baptist Church. Baptism by immersion, as shown here, is the customary way of signifying acceptance of Baptist beliefs.

be seen and not heard, and if by chance we talked or got caught chewing gum, Miss Nola or one of the ever-present ushers would take a long control stick and crack us on the head. The church rules were strict, and the ushers made sure nothing interfered with the high spirit of the service.

While the ushers proceeded to order the crowd, three of the deacons would place their chairs in front of the altar, for they were charged with starting the service. As I watched the activity of the church, my eyes fell on Mother Luella Byrd. Mother Byrd was not only the head of the Mother's Board, but basically in charge of the church. There she sat, dressed in white with her black cape draped over her shoulders, her arms folded and her face set. Once Mother Byrd had taken her position, God could begin to move....

Every Sunday morning, Mother Byrd was seated front and center at Saint Mark's by the time the singing began. As the song "I'll Fly Away" rang throughout the building, she rocked back and forth while the congregation rocked from side to side. While they sang, Elder Thomas began to preach. The singing and the preaching would blend and build together to a fever pitch. Elder Thomas, like an athlete at peak performance, paced the front of the church and preached until he was covered with sweat and the entire congregation was caught up in the spiritual fervor.

A national convention of members of the Nation of Islam in the 1960s. The Black Muslims emphasize the importance of black self-sufficiency and have established businesses in such cities as Detroit, Chicago, and New York.

A funeral in Los Angeles in 1932. This was an occasion when the mourners might sing some of the black spirituals, which greatly influenced American music.

Black churches in the South provided services that local white-dominated governments did not. In this church in Gee's Bend, Alabama, in 1937, a teacher uses a makeshift chalkboard to give adults and children a better education than they could receive in the segregated public school.

The pastor of a Pentecostal church in Chicago in the 1940s.

SCHOOLS

A segregated school in Alabama in 1945. Southern states provided little money to support what were supposed to be "separate but equal" schools for blacks.

Children in a kindergarten class in the 1890s in Kansas pick cotton from the garden they have planted.

Mary McLeod Bethune, born in 1875, became a leading African American educator. She described her own struggles to get an education.

I was born in Maysville, South Carolina, a country town in the midst of rice and cotton fields. My mother, father, and older brothers and sisters had been slaves until the Emancipation Proclamation....

[In those days] it was almost impossible for a Negro child, especially in the South, to get education. There were hundreds of square miles, sometimes entire states, without a single Negro school, and colored children were not allowed in public schools with white children. Mr. Lincoln had told our race we were free, but mentally we were still enslaved.

A knock on our door changed my life over-night. There stood a young woman, a colored missionary sent by the Northern Presbyterian Church to start a school near by. She asked my parents to send me. Every morning I picked up a little pail of milk and bread, and walked five miles to school; every afternoon, five miles home. But I walked always on winged feet.

The whole world opened to me when I learned to read. As soon as I understood something, I rushed back and taught it to the others at home....

By the time I was 15 I had taken every subject taught at our little school and could go no farther. Dissatisfied, because this taste of learning had aroused my appetite, I was forced to stay at home. Father's mule died—a major calamity—and he had to mortgage the farm to buy another. In those days, when a Negro mortgaged his property they never let him get out of debt.

I used to kneel in the cotton fields and pray that the door of opportunity should be opened to me once more, so that I might give to others whatever I might attain.

My prayers were answered. A white dressmaker, way off in Denver, Colorado, had become interested in the work of our little neighborhood school and had offered to pay for the higher education of some worthy girl. My teacher selected me....

[When I became a teacher] with the first money I earned I began to save in order to pay off Father's mortgage, which had hung over his head for ten years!

Rebecca Taylor was born in South Carolina around 1910. She described the experience of going to a segregated school and how it differed from the white schools.

I went to a black school. I think it was a four-room schoolhouse. There must have been two or three classes in each room. It was very crowded because sometimes while one class was being taught, we'd go outside. Then when we had class, we'd come inside and some of the others would go out. It was a frame building. Then there were schools for white children. The white schools were beautiful brick structures.

School segregation was maintained even when the number of African Americans was very small. Andrew Young, who would later serve as President Jimmy Carter's ambassador to the United Nations and as mayor of Atlanta, was born in New Orleans, Louisiana. He recalled growing up there.

I grew up in New Orleans in a neighborhood that was predominantly white. It was a poor neighborhood that was mostly Irish, Italians, Polish, and Cajuns who came in from rural areas, and my brother and I were the only black children in the neighborhood. So all of our early childhood playmates were white, and yet we went to segregated schools—we couldn't go to school with them. We were bused out of our neighborhood to go to an all-black school, and so it was almost like I grew up in two worlds. The society was segregated but we never accepted it. But neither did we resent it. I mean, my parents taught me this was not something to hate. They taught me that these were people that just didn't know any better and that even though we were all God's children, they just didn't believe that. They [simply] thought that white people were better than black people. But the Bible clearly says that "God is no respecter of persons." And so, I shouldn't let them get me upset because they didn't realize that I was God's child too. They just didn't know any better. So you have to help them to learn that.

So I went to a segregated public grammar school and to the Methodist High School. Our schools always had a strong Negro history component, and there was a lot of motivation to make the world a better place. We didn't think of it too much in terms of desegregation in those days. We thought of it in terms of being able to compete equally. We were taught that the world is structured against you because you are black and so you've got to work a lot harder than anybody white to succeed, but nobody ever thought of changing the situation. They just said, "In order to survive in the situation, you've got to be strong to put up with all you're going to have to put up with."

A first-grade class in Washington School in Topeka, Kansas. The lawsuit that overturned school segregation throughout the nation was filed against the Topeka school board by Oliver Brown on behalf of his 11-year-old daughter Linda.

In 1948, the Supreme Court ruled that the University of Oklahoma had to admit G. W. McLaurin, who became the university's first black student. In an attempt to maintain segregation, the university placed McLaurin in a separate room during his classes, as shown here.

Tennessee Town, Kansas, was one of the all-black towns founded by African Americans who left the South after the Civil War. The town was named after the state from which most of its residents came. Children and teachers pose here in front of the town's combination library, reading room, and kindergarten.

Tuskegee Institute sent "schools on wheels" like this to rural areas to teach the new methods of farming developed by Booker T. Washington.

Jake Lamar, who grew up in New York City in the 1960s, recalled his father's story of how he had escaped poverty in the South.

I'm an escapee from the garbage can," my father said proudly. "That's all I am. They tried to slam the lid down on me. 'Stay in there, boy! Stay down there with the rest of the trash!'... But I knew when I was a boy I wasn't stayin' in no Shingley, Georgia. Uh-uh. No way. 'Cause I was *smart*.... Now I look around me, sittin' up here in a six-room apartment in New York City, and sometimes it's like I can hardly believe it. I mean, I'm not *supposed* to be here...."

Dad launched into the story. "Fifteen years old and I was already a senior in high school, head of the damn class. Principal tells me he wants to see me in his office first thing Monday morning—and to bring my mother! So Monday morning Ma and I show up and the principal's looking all nervous and frightened when he answers the door. Sittin' there in his office is this fat, sweaty-faced cracker, says he's some kind of state education officer. And the cracker says to me, 'Boy—according to these here test scores, you the smartest nigra in the state.' And I say, 'Yessir, that's right.' And that cracker just stared at me for a long time and then he says, 'Boy—the state has decided that we want to send a nigra to college. And that nigra is you.' ... College! The white folks want to send me to college? And I look over at Ma and she's sittin' there lookin' like she's about to bust out crying. And then that flabby-assed cracker says, 'Boy—you can go to any college in the state. *Any college in the state*. Except...for the [white-only] University of Georgia, Georgia Tech, Georgia A & M, Emory...' And then Ma says, 'I want my son to go to Morehouse.' Which was just as well since the white folks weren't about to send me anyplace else."

Atlanta's Morehouse was regarded by many as the finest black college in America (some would rank Washington's Howard University higher, but none dared suggest this to my father). Dad excelled there, performing well in his courses, emerging as a popular student leader.... He also made his first acquaintance with the black bourgeoisie, the well-spoken, carefully tailored sons of ministers, lawyers, doctors and successful businessmen (including, two years ahead of my father, the son of Atlanta's most prominent clergyman, an unpretentious student named Martin Luther King, Jr.).

Mildred Arnold described her experience as the only black child in a school in Newark, New jersey, in the 1920s.

You had the children to fight because you were black. That was an everyday thing. And besides fighting the children, you had these little incidents with the teachers. I remember when I reached the fifth grade, the teacher was a Miss Messina. She stayed out of school the whole year so she wouldn't have to teach me. In those days one teacher taught all of the subjects to her class and she would have had me the entire year. She was determined that she wasn't teaching any black child. She wouldn't come to school till the day I got promoted out of the fifth grade. Then she came back to school to teach.

But the eminent poet Langston Hughes had a happier experience. He described the school he attended in Cleveland in the second decade of the 20th century.

I went to Central High School in Cleveland. We had a magazine called the *Belfry Owl*. I wrote poems for the *Belfry Owl*....

Little Negro dialect poems like Paul Laurence Dunbar's and poems without rhyme like Sandburg's were the first real poems I tried to write. I wrote about love, about the steel mills where my step-father worked, the slums where we lived, and the brown girls from the South, prancing up and down Central Avenue on a spring day....

My best pal in high school was a Polish boy named Sartur Andrzejewski. His parents lived in the steel mill district. His mother cooked wonderful cabbage in sweetened vinegar. His rosy-cheeked sisters were named Regina and Sabina. And the whole family had about them a quaint and kindly foreign air, bubbling with hospitality. They were devout Catholics, who lived well and were very jolly.

I had lots of Jewish friends, too, boys named Nathan and Sidney and Herman, and girls named Sonya and Bess and Leah. I went to my first symphony concert with a Jewish girl—for these children of foreign-born parents were more democratic than native white Americans, and less anti-Negro....

Four years at Central High School taught me many invaluable things. From Miss Diecke, I learnt that the only way to get a thing done is to start to do it, then keep on doing it, and finally you'll finish it, even if in the beginning you think you can't do it at all. From Miss Weimer I learnt that there are ways of saying or doing things, which may not be the currently approved ways, yet that can be very true and beautiful ways, that people will come to recognize as such in due time.

Henderson School in Fayetteville, Arkansas. It was founded in 1868 with aid from a national mission society and contributions from the citizens of Fayetteville. It was the first free school in the state.

As Others Saw Them

In 1991, Quinn Brisben, a white teacher in a black neighborhood in Chicago, told interviewer Studs Terkel that the aptitude tests used to estimate the potential of students were "horrible frauds."

Anybody who judges intelligence on a single number is a class enemy. The human mind is so great and diverse. My students know all kinds of wonderful things that they've had to learn in order to survive. But those things aren't on the tests. It's good to know Mozart, but they're never asked about B. B. King. You're supposed to know only a certain kind of grammar, a certain kind of vocabulary.

They tell you the way your parents spoke is bad and wrong. They say these are bad people who say "I be" or "I aks your mama." This simply isn't so. There is such a thing as standard English.... This is your ticket up and out. But what do you do when you go back to your mother and father and speak to them? I know all kinds of parents or grandparents who are uncomfortable among their educated children.

I think we should forge a bridge. You'll need standard English to communicate in the establishment. But no, your parents are not wrong. What your parents know is good to know, too. Matter of fact, the blues ought to be part of the curriculum.

The band of Sumner High School in Kansas City in 1918. Around that time, jazz music from New Orleans had reached Kansas City.

Spectators at New York City's first black rodeo in the 1960s. In that decade, the slogan "Black Is Beautiful" signified the renewed pride that African Americans took in their own tradition and styles.

CHAPTER SIX

"WE SHALL OVERCOME"

From the earliest days of colonization, African Americans have immeasurably enriched the nation with their labor, inventions, and art. They have fought for their country in every war from the Revolution on. One of black Americans' most cherished heroes, Joe Louis, the heavyweight champion of the world, gave up his boxing career to serve in the army during World War II.

African American veterans returning from the war found that conditions had not changed. In Georgia in 1946 a black World War II veteran was shot while attempting to vote, and his killer was acquitted by a white jury. That same year, race riots erupted in two southern cities.

There were some indications that African American citizens were slowly gaining victories in their struggle for full rights. In 1946, President Harry S. Truman created a Committee on Civil Rights, which published a report condemning racial injustices. But Truman could not persuade Congress—which was controlled by Republicans and southern Democrats—to pass even a limited civil-rights program. In 1948 Truman issued an executive order directing "equality of treatment and opportunity" in the armed forces. Even so, when the Korean War broke out in 1950, many African Americans still fought in segregated units.

The struggle to integrate the nation's schools had started long before. Two NAACP lawyers, Charles Hamilton Houston and Thurgood Marshall, succeeded in forcing the University of Maryland to admit a black student in 1935. Yet 15 years later, segregation was still the law or custom in most school systems. When nine-year-old Linda Brown was barred from attending an all-white public school in Topeka, Kansas, in 1950, the NAACP took up her case. Thurgood Marshall pursued it all the way to the Supreme Court.

In 1954, the court finally issued its decision in the case of *Brown* v. *Board of Education of Topeka, Kansas.* By unanimous decision, the court ruled in Linda Brown's favor. But the decision went much further than that: it overturned the 1896 decision in *Plessy* v. *Ferguson* that had been the basis for all segregated school systems. The following year the Court ordered state and local governments to proceed "with all deliberate speed" to integrate their schools.

More legal victories swiftly followed. In 1955, the Supreme Court also banned segregation in public recreational facilities, and the Interstate Commerce Commission ordered that buses and waiting rooms involved in interstate travel must be integrated.

That same year, one of the turning points in the civil rights struggle occurred in Montgomery, Alabama. On December 1, 1955, Rosa Parks, a member of the local NAACP, refused a bus driver's order to give up her seat to a white man. She was arrested and fined $14. In response, E. D. Nixon, head of the local NAACP chapter, and the Women's Political Council organized a boycott of the Montgomery bus system. Blacks started carpools and began walking instead of taking the bus. Among the leaders of the boycott was the young pastor of the Dexter Avenue Baptist Church, Martin Luther King, Jr. The boycott lasted for a little more than a year, and it forced the city to integrate the bus system.

By the time the boycott ended, King had gained national attention for his inspiring speeches and leadership. Soon after, he and Ralph D. Abernathy formed the Southern Christian Leadership Conference (SCLC), which began to carry the civil rights movement to other southern cities.

White racists did not willingly surrender to the decisions of the

Supreme Court. Throughout the South, White Citizens Councils were formed to resist integration. In 1957, a white mob prevented nine black students from entering Central High School in Little Rock, Arkansas. Television cameras carried the ugly scene into homes throughout the country, and President Dwight D. Eisenhower sent federal troops to Little Rock to enforce the law of the land.

King firmly believed that nonviolent actions such as boycotts and marches could arouse the conscience of the nation, and he was right. In 1963 the nation watched on TV as King led a protest march in Birmingham, Alabama. Using dogs, clubs, and fire hoses, the Birmingham police brutally attacked the marchers, arresting more than 2,000 people.

As King had hoped, Americans of all races responded with support for the civil rights cause. White and black students, along with priests, rabbis, ministers, and nuns, joined the growing protests in the South.

Other civil rights organizations took part in the fight. The Congress of Racial Equality (CORE), founded in Chicago in 1942, organized sit-ins in which blacks occupied seats in white-only restaurants until they were served (or, more often, evicted or arrested). Freedom Rides, in which blacks and whites sat together on buses traveling through the South, were another tactic used by CORE and the Student Nonviolent Coordinating Committee (SNCC).

In August 1963, the largest civil rights demonstration in the nation's history took place in Washington, D.C. More than 250,000 people listened as Martin Luther King, Jr., declared, "I have a dream that one day this nation will rise up and live out the true meaning of its creed: 'We hold these truths to be self-evident; that all men are created equal.'"

The civil rights movement gave African Americans new economic opportunities that enabled them to buy their own homes in suburban neighborhoods. Here, a backyard cookout at the home of the Gross family in California.

The following year, President Lyndon Johnson pushed through Congress the first Civil Rights Act since Reconstruction.

However, African Americans in northern cities were impatient for the civil rights movement to touch their lives as well. Riots and school boycotts broke out in black urban neighborhoods in New York, Chicago, and New Jersey in 1964. Northern blacks protested the racism that kept them from getting white-collar jobs, prevented them from buying houses in white neighborhoods, and barred their children from going to school in white districts.

African American leaders who did not share King's devotion to nonviolence began to speak out. After becoming head of SNCC, H. "Rap" Brown abandoned its nonviolent tactics. "Violence," he said, "is as American as apple pie." Malcolm X, an eloquent member of the Nation of Islam, proclaimed that blacks should take control of their own communities "by any means necessary." In 1965, Malcolm X was assassinated after leaving the Nation of Islam to start his own organization to work for black unity and freedom.

During the 1960s, black rage grew in the urban areas of the North. In 1968, major riots erupted in the Watts area of Los Angeles; Newark, New Jersey; and Detroit. King addressed the issue of urban poverty in March 1968 by announcing a "Poor People's Campaign" that would unite blacks, whites, and Latinos. A month later, King fell to an assassin's bullet.

But the civil rights movement could not be stopped, for it was the work of hundreds of thousands of people. In the 1970s, "affirmative action" plans required businesses and colleges to accept more African American applicants.

Slogans such as "Black Power" and "Black Is Beautiful" popularized the ideas that African Americans could influence society and no longer had to take a back seat in any area of American life.

In 1967, Carl Stokes became the first black mayor of a major city, Cleveland. The following year, nine African Americans were elected to Congress, the most in American history up to that date. Many more black elected leaders followed in their footsteps in towns, cities, and states throughout the nation. In 1988, Jesse Jackson won several state primary elections in his bid to become the Presidential nominee of the Democratic party.

African Americans came to be accepted as supervisors and managers in the nation's largest businesses. Their children began graduating from the nation's finest colleges, and for some the future looked bright. But as more prosperous blacks moved out of urban ghettos, those left behind sank deeper into poverty and despair. In the 1980s, the Reagan administration dismantled federal assistance programs, and drugs, crime, gangs, and homelessness threatened America's inner cities. Tragically, many young urban blacks feel that their only way out of the ghetto is to imitate the most prominent black Americans—those who are stars of sports and entertainment.

African Americans have excelled in these two areas because all they needed was a chance to display their talents. On the stage and the playing field, inherited wealth, social connections, and ties to the powerful cannot help a person succeed, as they can in business.

Before Jackie Robinson first took the field for the Brooklyn Dodgers in 1947, there were no African Americans in the major leagues of baseball. Blacks did not play in the National Football League between 1934 and 1946. Once the color barrier fell, African Americans' hard work and ability helped them excel in baseball, basketball, football, and tennis.

African American music has entranced the nation ever since a group of Fisk University students, the Jubilee Singers, went on a fund-raising tour in 1871. Spirituals, the distinctive religious songs of the slaves, were only the beginning of African American contributions to American art. In the early 20th century, the blues and jazz, with roots in African rhythms, swept up the Mississippi River to St. Louis and Chicago. Later, rhythm and blues emerged as a category of black music as urban blues performers began to incorporate jazz rhythms. Soul music in the late 1940s and early 1950s combined elements of rhythm and blues, black gospel music, and pop music. Rhythm and blues was also a major influence in the formation of rock music. Rap music, developed in black urban ghettos, is the latest manifestation of African American musical talent.

African Americans on stage, screen, and TV no longer have to take stereotyped roles. In 1977, Alex Haley's saga of an African American family, *Roots,* attracted the third-largest audience of any television show in history. Bill Cosby's TV show, depicting a middle-class black family, was popular all over the world. Oprah Winfrey, star of movies and TV, became wealthy by establishing her own production company.

African Americans have also excelled in literature and art. The painter Romare Bearden created hauntingly beautiful images of the rural life his family had left in the Great Migration. Jacob Lawrence painted series of panels that depict events in African American history. Gwendolyn Brooks, the first African American to win the Pulitzer Prize for poetry (1950), has been followed by such brilliant poets as Maya Angelou. The novelists Ralph Ellison and Richard Wright paved the way for Toni Morrison, who received the Nobel Prize for literature in 1993.

"The mystic spell of Africa is and ever was over all America," wrote W. E. B. Du Bois in 1909. "It has guided her hardest work, inspired her finest literature, and sung her sweetest songs." That statement remains true today.

Wearing the sign that he had marched with in civil rights demonstrations, a man pays tribute at the tomb of Martin Luther King, Jr.

THE CIVIL RIGHTS MOVEMENT

Jo Ann Robinson, an English teacher at Alabama State College who helped organize the Montgomery bus boycott, described the first day of the protest.

Monday morning, December the fifth, 1955, I shall never forget because many of us had not gone to bed that night. It was the day of the boycott. We had been up waiting for the first buses to pass to see if any riders were on them. It was a cold morning, cloudy, there was a threat of rain, and we were afraid that if it rained the people would get on the bus. But as the buses began to roll, and there were one or two on some of them, none on some of them, then we began to realize that the people were cooperating and that they were going to stay off the bus that first day.

Reverend Dr. Martin Luther King, Jr., was one of the leaders of the Montgomery bus boycott. On the first night of the boycott, King addressed 5,000 supporters.

There comes a time when people get tired. We are here this evening to say to those who have mistreated us so long that we are tired—tired of being segregated and humiliated, tired of being kicked about by the brutal feet of oppression.... If you will protest courageously and yet with dig-

More than 250,000 people gathered on the Mall in Washington, D.C., in 1963 to demonstrate in favor of the civil rights bill. Speaking to the crowd, Martin Luther King, Jr., said, "I have a dream that my four little children will one day live in a nation where they will not be judged by the color of their skin but by the content of their character."

nity and Christian love, in the history books that are written in future generations, historians will have to pause and say, "There lived a great people—a black people—who injected new meaning and dignity into the veins of civilization."

In September 1957, seventeen African American students were enrolled at Central High School in Little Rock, Arkansas. Despite the threat of mob violence, the superintendent of schools refused to allow a police escort for the children. Daisy Bates, the president of the state NAACP, asked a group of ministers to accompany the children to the school and called the parents to tell them where to assemble. However, Elizabeth Eckford was not contacted because her family did not have a telephone. As a result she went to school by herself, facing the hostile crowd. The governor had called out Arkansas National Guard troops who stood with bayonets drawn to "protect" the students. Eckford's account of her experience is found in Daisy Bates's 1962 autobiography.

That night [before the first day of school] I was so excited I couldn't sleep. The next morning I was about the first one up. While I was pressing my black and white dress—I had made it to wear on the first day of school—my little brother turned on the TV set. They started telling about a large crowd gathered at the school. The man on TV said he wondered if we were going to show up that morning. Mother called from the kitchen, where she was fixing breakfast, "Turn that TV off!" She was so upset and worried. I wanted to comfort her, so I said, "Mother, don't worry."

Dad was walking back and forth, from room to room, with a sad expression. He was chewing on his pipe and he had a cigar in his hand, but he didn't light either one. It would have been funny, only he was so nervous.

Before I left home Mother called us into the living-room. She said we should have a word of prayer. Then I caught the bus and got off a block from the school. I saw a large crowd of people standing across the street from the soldiers guarding Central. As I walked on, the crowd suddenly got very quiet....

I still wasn't too scared because all the time I kept thinking that the guards would protect me.

When I got right in front of the school, I went up to a guard.... He just looked straight ahead and didn't move to let me pass him. I didn't know what to do....

I stood looking at the school—it looked so big! Just then the guards let some white students go through.

The crowd was quiet. I guess they were waiting to see what was going to happen. When I was able to steady my knees, I walked up to the guard who had let the white students in. He too didn't move. When I tried to squeeze past him, he raised his bayonet and then the other guards closed in and they raised their bayonets.

They glared at me with a mean look and I was very frightened and didn't know what to do. I turned around and the crowd came toward me.

They moved closer and closer. Somebody started yelling,

In 1954, a mother and child sit on the steps of the Supreme Court with a newspaper reporting that the Court had struck down laws permitting racial segregation in public schools.

Elizabeth Eckford walks among hostile whites as she tries to enter Central High School in Little Rock in September 1957.

"Lynch her! Lynch her!"

I tried to see a friendly face somewhere in the mob—someone who maybe would help. I looked into the face of an old woman and it seemed a kind face, but when I looked at her again, she spat on me.

They came closer, shouting, "No nigger bitch is going to get in our school. Get out of here!"

I turned back to the guards but their faces told me I wouldn't get help from them. Then I looked down the block and saw a bench at the bus stop. I thought, "If I can only get there I will be safe." I don't know why the bench seemed a safe place to me, but I started walking toward it....

When I finally got there, I don't think I could have gone another step. I sat down and the mob crowded up and began shouting all over again. Someone hollered, "Drag her over to that tree! Let's take care of the nigger." Just then a white man sat down beside me, put his arm around me and patted my shoulder. He raised my chin and said, "Don't let them see you cry."

Then, a white lady—she was very nice—she came over to me on the bench. She spoke to me but I don't remember now what she said. She put me on the bus and sat next to me. She asked my name and tried to talk to me but I don't think I answered. I can't remember much about the bus ride, but the next thing I remember I was standing in front of the School for the Blind, where Mother works.... I kept running until I reached Mother's classroom.

Mother was standing at the window with her head bowed, but she must have sensed I was there because she turned around. She looked as if she had been crying, and I wanted to tell her I was all right. But I couldn't speak. She put her arms around me and I cried.

Martin Luther King, Jr. (second from left, front row) and other civil rights leaders—Whitney Young (sixth from left), Roy Wilkins (seventh from left), and A. Philip Randolph (eighth from left)—head the March on Washington in 1963.

The Freedom Rides organized by the Congress of Racial Equality (CORE) attracted people from all over the country to integrate the southern buses. Hank Thomas recalled a violent incident in Anniston, Alabama, on May 14, 1961.

The Freedom Ride didn't really get rough until we got down to the Deep South. Needless to say, Anniston, Alabama, I'm never gonna forget that. When I was on the bus they [whites] threw some kind of incendiary device on. I got real scared then. You know, I was thinking—I'm looking out the window there, and people are out there yelling and screaming. They [the mob] just about broke every window out of the bus.... I really thought that that was going to be the end of me. They shot the tires out, and the bus driver was forced to stop.... And we were trapped on the bus.

It wasn't until the thing [fire bomb] was shot on the bus and the bus caught afire that everything got out of control.... First they [the mob] closed the doors and wouldn't let us off. But then I'm pretty sure...that somebody said, "Hey, the bus is gonna explode...." and so they started scattering, and I guess

Fire hoses and dogs were used to disperse civil rights demonstrators in Birmingham, Alabama, in 1963. To stop the nonviolent sit-ins and marches, Birmingham's police arrested hundreds of demonstrators.

that's the way we got off the bus. Otherwise we probably all would have been succumbed by the smoke.... I got whacked over the head with a rock or I think a stick as I was coming off the bus.

The bus started exploding, and a lot of people were cut by flying glass.... Took us to the hospital, and it was incredible. The people at the hospital would not do anything for us. They would not. And I was saying "You're doctors, you're medical personnel." They wouldn't.... But strangely enough, even those bad things don't stick in my mind that much. Not that I'm full of love and goodwill for everybody in my heart, but I chalk it off to part of the things that I'm going to be able to sit on my front porch in my rocking chair and tell my young'uns about, my grandchildren about.

In December 1964, shortly before his assassination, Malcolm X spoke to African American teenagers from Mississippi who had come to New York City under the sponsorship of SNCC. He praised their efforts but added:

My experience has been that in many instances where you find Negroes talking about nonviolence, they are not nonviolent with each other, and they're not loving with each other, or forgiving with each other.... I think you understand what I mean. They are nonviolent with the enemy. A person can come to your home, and if he's white and wants to heap some kind of brutality on you, you're nonviolent; or he can come to take your father and put a rope around his neck, and you're nonviolent. But if another Negro just stomps his foot, you'll rumble with him in a minute. Which

Malcolm X addresses a crowd in Harlem in the early 1960s. In the two years before his assassination in 1965, Malcolm X broke his ties to the Nation of Islam. He made a pilgrimage to Mecca, the holy city of the Islamic religion, and changed his name to Al Hajj Malik el-Shabazz.

shows you that there's an inconsistency there.

I myself would go for nonviolence if it was consistent, if everybody is going to be nonviolent all the time. I'd say okay, let's get with it, we'll all be nonviolent. But I don't go along with any kind of nonviolence unless everybody's going to be nonviolent....

So we...are with the struggle in Mississippi to vote one thousand percent. We're with the efforts to register our people in Mississippi to vote one thousand percent. But we do not go along with anybody telling us to help nonviolently....

I hope you don't think I'm trying to incite you. Just look here: look at yourselves. Some of you are teenagers and students. How do you think I feel—and I belong to a generation ahead of you—how do you think I feel to have to tell you, "We, my generation, sat around like a knot on a wall while the whole world was fighting for its human rights—and you've got to be born into a society where you still have that fight." What did we do, who preceded you? I'll tell you what we did: nothing. And don't you make the same mistake we did....

So don't you run around here trying to make friends with somebody who's depriving you of your rights. They're not your friends, no, they're your enemies. Treat them like that and fight them, and you'll get your freedom; and after you get your freedom, your enemy will respect you. And we'll respect you. And I say that with no hate. I don't have any hate in me. I have no hate at all. I don't have any hate. I've got some sense. I'm not going to let somebody who hates me tell me to love him. I'm not that way-out. And you, young as you are, and because you start thinking, are not going to do it either.

On Sunday, March 7, 1965, in Selma, Alabama, a march was called to protest police brutality and the lack of voting rights for African Americans. Sheyann Webb was one of the marchers who faced the police on the Pettus Bridge on their way to the capital at Montgomery. She was only eight years old. She remembered the scene.

All I knew is I heard all this screaming and the people were turning and I saw this first part of the line running and stumbling back toward us. At that point, I was just off the bridge and on the side of the highway. And they came running and some of them were crying out and somebody yelled, "Oh, God, they're killing us!" I think I just froze then. There were people everywhere, jamming against me, pushing against me. Then, all of a sudden, it stopped and everyone got down on their knees, and I did too, and somebody was saying for us to pray. But there was so much excitement it never got started, because everybody was talking and they were scared and we didn't know what was happening or what was going to happen.... It seemed like just a few seconds went by and I heard a shout, "Gas! Gas!" And everybody started screaming again. And I looked and I saw the troopers charging us again and some of them were swinging their arms

In 1968, the governor of Tennessee called out the National Guard to block protest marchers in Memphis. The marchers carry signs with a message signifying that they were no longer contented to be second-class citizens.

and throwing canisters of tear gas. And beyond them I saw the horsemen starting their charge toward us. I was terrified. What happened then is something I'll never forget as long as I live. Never. In fact, I still dream about it sometimes.

I saw those horsemen coming toward me and they had those awful masks on; they rode right through the cloud of tear gas. Some of them had clubs, others had ropes or whips, which they swung about them like they were driving cattle.

I'll tell you, I forgot about praying, and I just turned and ran. And just as I was turning the tear gas got me; it burned my nose first and then got my eyes. I was blinded by the tears. So I began running and not seeing where I was going. I remember being scared that I might fall over the railing and into the water. I don't know if I was screaming or not, but everyone else was. People were running and falling and ducking and you could hear the horses' hooves on the pavement and you'd hear people scream and hear the whips swishing and you'd hear them striking the people. They'd cry out; some moaned. Women as well as men were getting hit. I never got hit, but one of the horses went right by me and I heard the swish sound as the whip went over my head and cracked some man across the back. It seemed to take forever to get across the bridge. It seemed I was running uphill for an awfully long time. They kept rolling canisters of tear gas on the ground, so it would rise up quickly. It was making me sick. I heard more horses and I turned back and saw two of them and the riders were leaning over to one side. It was like a nightmare seeing it through the tears. I just knew then that I was going to die, that those horses were going to trample me. So I kind of knelt down and held my hands up over my head, and I must have been screaming—I don't really remember.

All of a sudden somebody was grabbing me under the arms and lifting me up and running.... And I looked up and saw it was Hosea Williams who had me...and I shouted at him, "Put me down! You can't run fast enough with me!"

But he held on until we were off the bridge and down on Broad Street and he let me go. I didn't stop running until I got home.... I remember just laying there on the couch, crying and feeling so disgusted. They had beaten us like we were slaves.

During the summer of 1967, the pent-up rage of African Americans exploded. Riots broke out in almost 150 American towns and cities. Hundreds of people were killed and hundreds of millions of dollars worth of property destroyed. The worst riot of that bloody time started in Detroit on July 23, and lasted for five days. Federal troops had to be called in to restore order. Ron Scott, a 20-year-old auto worker, related one experience of those terrifying days.

We lived on the 14th floor of a 14-story building in the Jeffries Housing Projects. One night...in the middle of the rebellion, my mother, my five-year-old sister, and my three-year-old brother were there.... A neighbor of ours apparently had brought a rifle into the house and we heard sev-

Black students do their homework while taking part in a sit-in at a Woolworth's lunch counter in Little Rock, Arkansas, in 1962.

In the 1960s, police in Atlanta, Georgia, arrest a civil rights demonstrator. Only a few years later, in 1973, Atlanta elected a black mayor, Maynard Jackson, for the first time in its history.

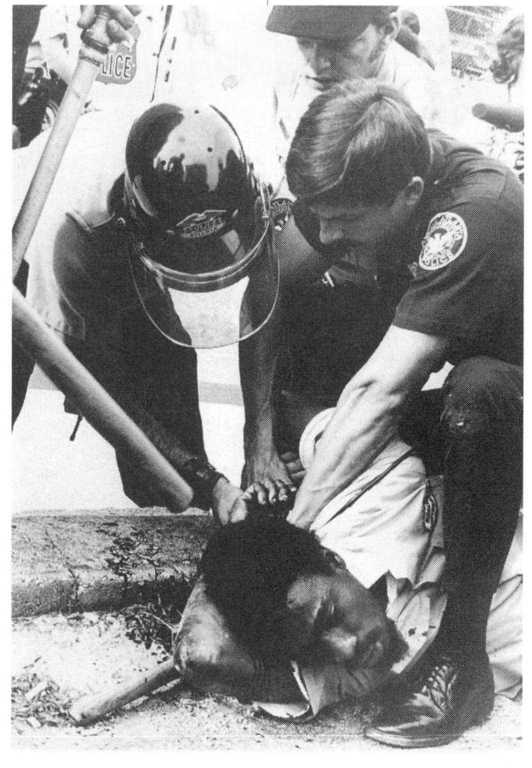

eral shots. I was looking out of the window trying to see where the shots were coming from. And the next thing we know, there is a blast of machine-gun fire coming past the building. We fall on the floor and turn out the lights. Next thing we heard, about two, three minutes later, was this pounding, pounding, pounding, pounding on the door. They were knocking like they were gonna cave the door in.

On my side of the door, I'm standing there wondering exactly what's gonna happen when I open that door. When I open the door, there are about four or five National Guardsmen. All young, all white, looking around, with rifles and bayonets. They come in our house, in our living room, they're standing there. And by this time, in the three, four days of the rebellion, there's been people killed. There's been people shot on the street, for no reason whatsoever. By this time, I'm angry, I'm fearful of what's happening. And this one guy says, "We heard some shooting here."

A march by the Black Panthers, a militant civil rights group founded in Oakland, California, in 1966 by Bobby Seale and Huey Newton. At the time of this march in 1968, Newton had been charged with the murder of a policeman. He was found guilty of manslaughter, but the California Supreme Court later overturned his conviction.

And I said, "There was no shooting here."

He says, "Yeah, we heard some shooting here." And this guy is standing looking at me—at any moment he can blow my head off, my sister's, and my brother's.... And I knew that if I was shot, if my family was shot, that they could have closed the door on this apartment and nobody would have ever known what happened. When I looked in his eyes, and he looked at me, it looked to me as if he wanted to kill somebody.

And just then one of the other guardsmen said, "The shooting came from down here." They ran down to the guy's apartment where the shooting had taken place...[and] they dragged him on the elevator.

When I think about the fact that some guy, from outside of Detroit, who didn't even know us, could've blown us away, it makes me mad. It makes me...realize that as my mother says, God was protecting us. I believe that if the guy who came in and said that the shooting was happening down the hall hadn't come in, we might not have made it. There wasn't anything I could do.

On April 4, 1968, Dr. Martin Luther King, Jr., was killed by a sniper while he stood on the balcony of a motel in Memphis, Tennessee. Ralph Abernathy, King's close friend, was standing nearby, and described the scene.

I heard what sounded like a firecracker. And I jumped. And when I jumped I saw only his feet laying on the balcony. And I immediately rushed to his side and I started patting his cheek, saying, "Martin, Martin, Martin. Don't be afraid. Don't be afraid. This is Ralph. This is Ralph. This is Ralph." And I got his attention. And he calmed down. His eyes were moving and he became very, very calm....

[When an ambulance arrived,] I went with him, rode with him in the back of the ambulance. And I...would not leave the operating room. And finally the doctor came over to me and said, "You are Dr. Abernathy? He will not survive. It will be an act of mercy, because he would be paralyzed from his waist down. You may have your last moments with him."

And I went over and took him in my arms. And he breathed his last breath.

Luke Harris attended St. Joseph's College in Philadelphia and received a law degree from Yale University in 1977. But as he explains, his educational achievements would not have been possible without affirmative action.

I'm deeply interested in affirmative action because I'm a product of it. I was fortunate enough to be raised by a very loving great-aunt and a great-uncle who...gave me everything that you could have expected and more. But...I spent the first six or seven years of my education in a segregated elementary school in Merchantville, New Jersey.... Now, by the time I was in ninth grade, I was getting signals from the guidance counselors that, you know, college is not for you. You're not the kind of guy that's every going to learn how to do things like chemistry and calculus and physics.... Some of it was a little bit disturbing....

Without affirmative action, there is no doubt that I would not have been able to go to St. Joe's. I worked very hard and I wound up graduating number one in my department, and that's when I wound up with the opportunity to go to Yale Law School. So I went to Yale Law School, feeling that I was part of the crest of a social movement. And that American society was finally opening up in some limited ways to allow people of color and blacks in particular to participate in all aspects of American life. And this was a first-time kind of thing. It had never happened before in America. And I felt proud and I still do feel proud to be a part of that process.

Birdie Lee May, born in Arkansas in 1904, grew up in a segregated society. She commented to an interviewer in the 1970s about the changes brought by the civil rights movement.

I can see things changing for the better. I was told when I was growing up, "You'll be thankful for this and that," but I thought, "I'll never be thankful for this kind of stuff!" I have it much easier now. Negroes, they have all my life felt inferior. Now the tide is changing and the other races don't mind this feeling too much and we have a few more advantages today. In the past I couldn't go to either restaurant in town. Negroes had a little nasty place. But I'm free now to go to any restaurant if I have the money to pay for it. Even to the best motels I can go to. Quite a few differences. There's a difference in schools. They used to say, "Equal but separate." That wasn't too long ago. Changing the schools helped.

Jesse Jackson speaks to reporters at the 1972 National Black Political Convention in Gary, Indiana. In 1984 and 1988, Jackson ran for the Democratic party's Presidential nomination. Though he lost, his success in attracting a sizable number of white voters has made him one of the most respected black American leaders.

Muhammad Ali, the only man to capture the heavyweight boxing championship three times, was also known for his speaking style and witty poems. Here, he speaks to a convention of the Nation of Islam in 1968. Seated behind him is Elijah Muhammad, who led the Black Muslims until his death in 1975.

In 1993, after leading the Chicago Bulls to three straight NBA championships, Michael Jordan quit to play pro baseball.

SPORTS

Joe Louis was heavyweight boxing champion of the world from 1937 to 1949. Throughout the nation, blacks felt pride in the achievements of this sharecropper's son. Clifton L. Taulbert described Joe Louis's importance in the small town of Glen Allan, Mississippi.

So few were the colored heroes that the ones we knew by name such as Joe Louis were like the neighbors who lived next door or down the road—a good ol' boy from the plantation that had gone north and made us proud....

I don't remember the exact year, but I recall the night our front yard was filled with people who had come over...to listen to the fight.... There weren't very many radios in the colored section, but Ma Ponk had an old battery-operated one. It was important to her. She played it only on Sundays and for special occasions, and this was a special occasion.

A crowd of people started gathering early.... Their hard day's work wasn't slowing them down—not on this night. Their hero would be fighting tonight, and Joe Louis with his fists, quickness and punching power could say for them what they could never say for themselves....

At eight o'clock someone mildly said, "It's time, Miss Ponk." She hooked the antenna wire into the screen and turned the knob. With a small amount of static, the fight was on. As the crowd listened, ignoring the bites of mosquitoes and night bugs, Joe Louis defended his title.

I remember the men jumping up and down, shadowboxing with each other. They gave step-by-step instructions, as if they were coaches and Joe Louis could hear each word....

The radio fight would go on into the night, too late for a little boy like me to hear it to the finish. I'd fall asleep on my cot by the window, while the colored people of Glen Allan successfully coached their hero to another victory.

In 1947, Branch Rickey, general manager of the Brooklyn Dodgers, selected Jackie Robinson to be the first African American player in major league baseball. Rickey warned Jackie that he would have to endure racial taunts. For the intensely proud young man, that was the greatest trial he had to face. Robinson recalled that season.

The first racial "incident" occurred in April, when the Phillies came to Brooklyn for a three-game series. The Phillies...are great bench-riders. The first time I stepped up to the plate, they opened up full blast. "Hey, you black Nigger," I heard one of them yell. "Why don't you go back where you came from?" Then I heard another one shout: "Yeah, pretty soon you'll want to eat and sleep with white ball players!" As the jockeying continued on this level, I almost lost my head. I started to drop my bat and go over and take a sock at one of them. But then I remembered Branch Rickey's warning me of what I'd have to take without losing my temper. So I pretended I didn't hear them. I gritted my teeth and vented some of my anger on a solid single.

On April 8, 1974, Hank Aaron of the Atlanta Braves hit his 715th home run, breaking the lifetime record long held by Babe Ruth. Aaron described the moment he faced Los Angeles Dodger pitcher Al Downing.

Downing...threw his slider low and down the middle, which was not where he wanted it but which was fine with me. I hit it squarely, though not well enough that I knew it was gone....

[Outfielder Bill] Buckner...ran back to the wall and turned. But the ball kept going. It surprised him, and it surprised me.... Anyway, something carried the ball into the bullpen, and about the time I got to first base I realized that I was the all-time home run king of baseball.... It was like I was running in a bubble and I could see all these people jumping up and down and waving their arms in slow motion. I remember that every base seemed crowded, like there were all these people I had to get through to make it to home plate. I just couldn't wait to get there. I was told I had a big smile on my face as I came around third. I purposely never smiled as I ran the bases after a home run, but I suppose I couldn't help it that time.

Years after becoming a basketball star for the Los Angeles Lakers, Kareem Abdul-Jabbar returned to Inwood, the New York City neighborhood where he grew up. The trip brought back memories.

We drove to Nagle Avenue, where I lived in Building Five of the Dyckman Street projects. It was a two-bedroom apartment up on the fifth floor, and my parents had one bedroom and I had the other. This was middle-income housing, city-owned.... You used to be fined $2 for walking on the grass. The subway, which was elevated in Inwood, ran right down Nagle Avenue, right by our building. I remember its sounds, at all times of day or night....

[Abdul-Jabbar stopped by St. Jude's elementary school, which he had attended.] I walked into the small front lobby, the chapel to my left, the doors to the gym straight ahead. I stepped inside. It was exactly the same. I remembered *all* of it. The baskets at either end of the court, the wooden backboards, the letters SJ in the circle at center court. I looked down to the basket where I first leapt toward the hoop and dunked a basketball.

I was thirteen. It was during a game early in the eighth grade and our guard Patrick Dourish flipped me a real nice pass on the fast break. The ball came into my hands and I was able to jam it. I think I was as shocked as everybody else. A moment of elation on the court I have never forgotten.

At the 1988 Olympics, Florence Griffith Joyner won three gold medals and one silver in track events—more than any other American female track athlete had ever won. Here, she crosses the finish line in the 100-meter dash.

Arthur Ashe was one of the few black athletes to excel in the sport of tennis. He became the first African American to win the men's singles title in the U.S. and British championships. In 1992, Ashe announced that he had contracted AIDS from a blood transfusion; he died in 1993.

ENTERTAINMENT AND MEDIA

Until the 1960s, there were few roles for blacks in Hollywood movies, except as servants. However, "race" movies with all-black casts were produced for African American audiences. Here, some of the actors in such a movie pose for a publicity shot in 1930.

As African Americans left the South, they took their music with them. From New Orleans and towns along the Mississippi up to Chicago and Kansas City, jazz, blues, and gospel became part of the nation's music. Muddy Waters (born McKinley Morganfield) was a blues musician from the Mississippi Delta who went to Chicago in 1943. In an interview, he described the thrill of his first hit record.

I think they released [the record] on a Friday. By noon. He pressed up 3,000 and delivered 'em and you couldn't buy one Saturday evening, you couldn't get one in Chicago nowhere. They sold 'em out, the people buyin' two or three at a time.... The next week...they had 'em pressin' records all night. They started a limit, one to a customer.... Hey, I had worked all my life for that. I wasn't thinkin' about money part of it that much, because I had worked all my life to get my name up there, and that was the truth.... I had a convertible, you know. I had the top back on it, and at night after I'd got through playing I'd be going home I could hear 'em—that record all up in people's houses. And I'd stop my car and look up and listen a little while. *Ooooh,* once I got a little scared of somethin', "What's gonna happen to me, you know?" All of a sudden I became Muddy Waters. You know? Just over night. People started to speakin' hollerin' across the streets at me.

Countless black musicians and singers have won nationwide popularity since the early years of this century. The recordings of blues singer Billie Holiday (1915–59) preserve the haunting quality of her voice.

John H. Johnson began his publishing empire in 1942 with the magazine Negro Digest. *He parlayed a $500 loan on his mother's furniture into the largest black-owned corporation in the country. Forty-five years later he remembered starting the magazine.*

November 1, 1942. I remember that Sunday as if it were yesterday. I was twenty-four years old, a recent migrant from Arkansas, lost and found in the big city of Chicago....

I'd lived all my life on the edge of poverty and humiliation. People say a lot of foolish things today about poverty. But I *know*—I've been there. It's not so much pain in the belly; it's pain in the soul. It's the wanting and not having, the eyes that see you and don't see you, and, all the while, just out of reach, on the other side of the glass bars of your cell, the sweets, the lights, the goodies, and the somebodyness.

I tell you, I *know*—I've been there and back.

In 1942, my mother and I were recent graduates of the relief rolls. And I'd decided that I was never going down that road again—*never*. I was willing to go anywhere. I was willing to do anything, or almost anything, to get some of the good things of this life.

You understand what I'm saying? I'm saying that *I had decided once and for all. I was going to make it, or die.*

The new magazine was a down payment on that dream. It was my stake at the big table of people who made deals and drove big cars and lived on Lake Shore Drive and migrated to Palm Springs when Lake Michigan froze over.

Would the new magazine take me there?...

Most important of all, and most frightening of all: *Would I disgrace my mother before her friends?*

That was the question.

Would the sheriff take my mother's furniture and put us out on the cold and mean streets?

The odds against me on that day were at least $200 million—that's what people *say* I'm worth—to 1. And that's not counting the handicap of race.

But I beat the odds.

I proved that long shots *do* come in.

The magazine I published on that faraway November day opened a vein of pure black gold. And in tribute to the god of November I made November my signature month.

On November 1, 1945, I started *Ebony.*

On November 1, 1951, I started *Jet.*

In other Novembers, in other years, I founded Fashion Fair Cosmetics and Supreme Beauty Products, and bought controlling shares in Supreme Life Insurance Company, where I started my career an office boy and a gofer.

Today I own the biggest Black-oriented corporation in America, and I sit on the boards of some of America's biggest corporations.

What a difference forty-five years make!

"The Cosby Show," a 1980s TV show about a middle-class black family (cast shown here), became a favorite of audiences all over the world. Its creator and star, Bill Cosby, is one of the most popular American entertainers. In the back row, second from right, is Phylicia Rashad, who played Cosby's wife in the show.

By the time Spike Lee was 35 years old, he had directed five feature-length motion pictures—a record that few filmmakers of any race can match. Each of his films has been controversial because Lee consistently attacks "conventional" ideas of race and of black-white relationships. Spike Lee has formed his own production company and established a fellowship for minority filmmakers at New York University. He encourages others to follow in his footsteps.

To tell you the truth I hate talking about my work, but I'd rather tell my story than have somebody else do it. Growing up I wanted to be an athlete. The sport didn't matter; it just depended on what season it was: basketball, football, baseball, I played 'em all and still love 'em today. I had no idea that people made movies. I just didn't know. You went to the movie house, the lights went out, the movie came on, you enjoyed it, you ate as much popcorn and candy as you could eat.... Movies were magic—and something you couldn't do. Or so you thought.

It's this perception of movies (which Hollywood promotes) that keeps folks from becoming filmmakers. We've been fed this hocus-pocus BS, so you think you can't do it. Filmmaking

Spike Lee has written, directed, and acted in most of his popular movies. Shown here in the editing room, Lee insists on controlling every aspect of production to express his personal vision on the screen.

The Apollo Theater

The Apollo Theater, on 125th Street in Harlem, became a showcase for African American talent. Among its most popular shows was the weekly amateur competition, at which the audience decided the winner. Many of the greatest African American musical performers, such as Ella Fitzgerald, James Brown, and Dionne Warwick, as well as comedian Richard Pryor, got their start by winning the Apollo contest. Francis "Doll" Thomas, who operated the lights at the Apollo, told an interviewer:

All the performers wanted to come here. If you hadn't played the Apollo, you hadn't been anywhere. The Apollo happened to have one of those peculiar types of audiences that appreciated a performer's talent and ability.

Here, they heard colored performers sing. They saw colored people dancin'. They saw a line of colored girls that were beautiful and talented, so naturally they appreciated it, that all. You appreciate what your people can do.

We were successful with the Apollo because we gave people what show business was supposed to be—that's entertainment. You could be down in the dumps, a little sick or not feelin' good, and I guarantee you, if you came to the Apollo and spent 75 or 80 minutes with our show, when you walked out, you were like a rejuvenated person.

is a craft, and it can be learned like anything else; of course, it takes talent, but forget about it being something magical and mystical. There is a reason for that party line: Film is a powerful medium; it can influence how millions of people think, walk, talk, even live, plus you can make enormous sums of money. The idea is to keep the industry confined, let a small group of people have the control and make all the money. This is why one of my goals has been the demystification of film. I like to tell and show people it can be done. I'm saying don't fall for junk like, *You gotta be struck by lightning to be a filmmaker.*

Naomi Waller Washington was the sister of Thomas "Fats" Waller, a renowned jazz pianist. Naomi Washington told how her brother became interested in jazz.

Fats didn't start playing jazz until he met Jimmy Johnson. James P. was the best, and Fats followed him foot by foot. Sometimes James P. would come over to Fats' house. Fats would be playing, and James would say, "No. Get up. I'll show you how to do it."

Just jabbing with one another. Naturally, James P. would sit down, and he'd play his version.

"Aha," Fats said. "Now you get up."

And they would carry on like that....

I loved Fats' playing. He used to relieve this woman at the Lincoln [a theater where silent movies were shown with live musical accompaniment], and he'd play the music for the picture. Then they got this organ. Now Fats took over that organ. You play as many notes with your foot as you do with your hands, and he had it all down after six lessons....

My mother was a very good singer. We all more or less liked to sing, and we all liked to sing while he played. The neighbors liked to hear Fats play the piano. Instead of shouting, "Stop that music," they'd say, "Raise that window, let us hear."

Helen Brown, a concert pianist and singer, told an interviewer that her career began when she sang "spirituals," the traditional religious music of southern African Americans.

I was with the Hall Johnson Choir, which was formed primarily to perpetuate Negro spirituals.... The spirituals were looked down upon by blacks and whites. People didn't think there was any particular quality to them. They didn't know they were the soul of the black man. They claimed that their words and their language weren't "ultra-ultra." They were in very bad English. They didn't know they were better English because they had more meaning....

Because of the choir, more and more people started to sing them, until everybody was trying to sing them. We performed with the New York Philharmonic Orchestra with Toscanini three times. To see a group of Negro singers with the New York Philharmonic was a feat and a curiosity. It should have been. We were people with voices. We sang operas. People were surprised that we could sing other things besides spirituals.

Howard "Stretch" Johnson was one of the dancers who performed at the Cotton Club. Many years later he taught black studies at the State University of New York at New Paltz and edited the first African American newspaper in Hawaii. Johnson recalled hearing Edward "Duke" Ellington play in the late 1920s.

I had always been a great admirer of Duke Ellington's. As a twelve-year-old kid, I used to listen to him on our radio when they first opened up the Cotton Club....

Ellington was a genius. Like most geniuses, he had a certain kind of sense of self, a charisma, and also he was a very kind person. He exuded warmth to the musicians in the band and to the other performers. He was almost like a big brother or father figure to most of us. We just idolized Duke. He was really a prince of a person, with what...later came to be called "cool" during the bop [music] period. Cool was already a part of the black ethic in those days. It was a kind of laid-back quality, taking things with grace, not being flustered or uptight, a kind of attitude toward life that the right is going to win out.

Diana Ross, one of the most popular African American singers ever, described how she, Mary Wilson, and Florence Ballard formed a singing group, later known as the Supremes. Diana was 14 when her family moved to the Brewster housing projects in Detroit.

It was there...that I met Mary Wilson and Florence Ballard.... We loved to sing more than anything, and we used to get together and rehearse as often as we could, with no idea how to get to the next step. When you want something bad enough, somehow it happens....

I finally decided to ask Smokey Robinson if he could get us an audition with Berry Gordy at Motown....

We stood up there in a straight line in front of some Motown representatives to do our audition, our nerves on edge, our emotions highly amplified. They were all strangers to us. Smokey wasn't in the room; he was waiting outside. We sang a few songs a cappella that we had painstakingly prepared for this day.... Then, in the middle of the last one, "There Goes My Baby," a song in which I sang lead, Berry Gordy himself came walking into the room.... Our movements got bigger, and our voices became more expressive. He stood there until we finished the song and he said, "Sing that song over again, 'There Goes My Baby.'" So we did. "Very nice, girls. Very nice." That was all he said, and then he swiftly left the room....

When he finally came back, he was somewhat encouraging. He liked us, he liked our sound, but he insisted that we were too young for a contract.

"Go and finish school, work on your music, and then come back and audition again."

Toni Morrison

In October 1993, Toni Morrison became the first African American to win the Nobel Prize for literature, the most coveted award for a writer. Throughout her career as a novelist, Morrison has eloquently portrayed the lives of black men and women, examining the influence of the past on the present.

Born in 1931 in Lorraine, Ohio, Morrison was the daughter of George and Ramah Wofford. Her father had moved north from Georgia, where racial violence had embittered him. Morrison's childhood was a peaceful one, in which she learned to love music and listened to tales of black folklore from her grandmother, who decoded dreams from a book of symbols.

Morrison started writing in the late 1950s, when she was a teacher of literature at Howard University. A short story she wrote at that time, based on a childhood memory of a neighbor girl who prayed that her eyes would turn blue, was the basis of her first novel, *The Bluest Eye,* published in 1970.

Financial rewards for her works came slowly, and Morrison supported herself as a teacher and editor. In these roles, she influenced other notable black writers, including Claude Brown, Toni Cade Bambara, and Alice Walker.

Her fifth novel, *Beloved* (1987), was a best-seller that won the Pulitzer Prize. In *Beloved,* Morrison relates the true story of an escaped slave who kills her daughter to keep her from being recaptured. Morrison dedicated it to the "sixty million and more" who endured the horrors of slavery.

Jazz, the novel she published just before winning the Nobel Prize, concerns another murder, in Harlem in the 1920s. In this book Morrison examines the anguish of those, like her father, who left the rural South during the Great Migration. Morrison has said that she writes for her own satisfaction, trying to discover the meaning of her personal experiences and those of her people.

Hoe Cakes

Vertamae Smart-Grosvenor, who is today a writer in New York City, grew up on a sharecropper's farm in South Carolina. In her cookbook-autobiography, Vibration Cooking: Or the Travel Notes of a Geechee Girl, *Smart-Grosvenor describes hoe cakes made by her grandmother.*

My Grandmother Sula was very beautiful. She was a Myers before she married Granddaddy Ritter. The Myerses got a lot of Indian blood and it sure showed on Grandmama Sula. She had high cheekbones and long black hair. Mostly I remember her sitting on the porch of the house that Granddaddy built when they first got married, smoking her pipe and rocking in her favorite chair.... I loved Grandmama Sula cause she was so softhearted and kind to us. She let us do things that she never let her children do or at least that's what Mama says. She could cook and she would sometimes cook Indian food for us. Mostly I remember the hoe cakes she made.

INGREDIENTS

2 handfuls sifted corn meal
pinch salt
pinch [baking] soda and melted bacon grease
(if you eat swine—peanut oil if you don't)
about 1 cup sweet milk

Mix all together and wait a few minutes to see if more water is needed—then pour the whole mixture into a hot greased heavy black cast-iron skillet. Put it on your plate. Pour some thick syrup and sop it up and it's out of sight. Hoe cakes are also good with pot likker [the juice left after boiling vegetables with meat]. Hoe cake got its name from the hoe. Slaves would cook batter on the flat edge of the hoe in the fields for the noonday meal. You don't have to cook it on the metal part of the hoe cause we ain't slaves no mo'.

A Kwanzaa celebration in the Crispus Attucks Center in Brooklyn, New York, in 1991. The seven candles stand for the seven principles of Kwanzaa.

CELEBRATING THE HERITAGE

Many African Americans today celebrate their heritage and roots by observing the seven-day festival of Kwanzaa. On each day, a family member lights a candle and discusses one of the seven principles of Kwanzaa—unity, self-determination, collective work and responsibility, cooperative economics, purpose, creativity, and faith. Eric Copage describes his family's Kwanzaa observance.

I was never a holiday kind of guy. Perhaps it was because we observed few holiday rituals of any kind. Although we put up a Christmas tree every year, there was no ceremony to it—no drinking of eggnog or listening to carols while hanging ornaments. To me the tree seemed more or less like another piece of furniture. Over the past few years, however, the holiday season has taken on a new meaning for me as my family sits at the dinner table the week following Christmas to celebrate Kwanzaa.

This cultural observance for black Americans and others of African descent was created in 1966 by Maulana (Ron) Karenga, who is currently chairman of black studies at California State University in Long Beach. Kwanzaa means "first fruits of the harvest" in Swahili, but there is no festival of that name in any African society. Karenga chose Swahili, the lingua franca of much of East Africa, to emphasize that black Americans come from many parts of Africa. Karenga synthesized elements from many African harvest festivals to create a unique celebration that is now observed in some way by more than 5 million Americans.

When I first told my wife I was thinking about observing Kwanzaa, she barred the way to our attic and said she'd never chuck our Christmas tree lights and antique ornaments. I told her that wouldn't be necessary. Kwanzaa, which runs from December 26 to New Year's Day, does not replace Christmas and is not a religious holiday. (We now celebrate both.) It is a time to focus on Africa and African-inspired culture and to reinforce a value system that goes back for generations....

When my family lights the black, red, and green Kwanzaa candles the last week of December, we do so with millions of other black Americans around the nation. Major community celebrations are held in just about every city that has any kind of black population.

Salim Muwakkil, who works for a magazine in Chicago, described his feelings about his African heritage.

One of the reasons I changed my name [to a Muslim one] is because, while in the service, I was stationed not far from Macon. My maternal family was there, so I made quite a few visits. They told me lots of stories. One was about an ancestor called Guinea Sultan. He came as a slave and was part of my family's history. It was the only connection I could find between myself and the land of my origin. So I feel a soft spot at least for Islam.

When this campaign to call ourselves African-American began, I was astonished by the hostility of so many white people to the idea. Why should they be? Africa is our homeland and we're American. We were always taught there was something wrong with who we were. If we openly accept our African-ness, is that subversive of American values? It's curious that white people want us to abandon our African heritage, which we've never been able to embrace.

When we first came here, it was against the law to practice our rituals, to recognize our heritage. Even the drum was forbidden. So now, when black people are just beginning to embrace their ethnic identity, whites look at it as some sort of threat.... It's simply that black people are just beginning to realize how much they've been deprived of knowledge of their own selves.

I think we must begin to realize that we are Americans of African descent. And to be comfortable with that identity. I've been to Africa several times, and to other countries. I know that I am peculiarly American. There is no getting around that.

Today's African Americans, like this family at an African festival at Boys and Girls High School in New York City, teach their children to take pride in their heritage.

Children at the African American Day Parade in Harlem, held each year in September.

Herman "Skip" Mason holds a framed picture of his great-great-aunt on his father's side. Behind him are some of the numerous pictures of his family members that he has collected over the years.

THE MASON FAMILY

Herman "Skip" Mason's home in Atlanta, Georgia, reflects his lifelong interest in the history of his family. Framed photographs of relatives cover the walls. His vast collection of pictures includes historical prints, pictures of Negro league baseball players, and photos of African Americans in every possible occupation, from a farmer to a group of surgeons in an operating room. One of his most prized possessions hangs over his living-room sofa: a photograph of his great-great-grandmother, born in 1834.

What started as a boyhood hobby has now become a business, called Digging It Up. Skip Mason helps others trace their family history and provides historical pictures of African Americans for TV shows, documentaries, movies, magazines, and newspapers. He also teaches African American history at Morehouse College, the alma mater of Martin Luther King, Jr.

Q: How did you first become interested in researching your family's history?

In 1977, when I was 14, Alex Haley's book *Roots* was made into a television series. I was one of the millions who watched *Roots* and I was overwhelmed by what I saw. I had very little background in the history of my people, African Americans. Because it really wasn't taught in the classroom. And I saw *Roots* and it made me realize that I had a larger family.

I started questioning my mother, who gave me some good information, and I began writing letters to relatives in Florida. Then I began to call them on the telephone. And our family phone bill grew to $200 or $300 a month, because I had become a fanatic. I started asking them questions about people that I did not know about, such as my great-grandparents. All of a sudden my family became real to me. You know, they were people I was descended from, whom I was a part of, but I knew very little about. But I could connect with them, because these were people whose lives were important way before mine. And because of their existence, their survival in life...I'm here today. Sometimes I think, what if just one of them would have just given up, or had died. There's a possibility I would not be here today.

This precious photograph of Mason's great-great-great-grandmother Ellen Barton was taken in the 1890s. She was born a slave in 1834.

Q: You have traced your family roots back quite a long way.

On my mother's side, I was able to trace my ancestry back to my fourth great-grandmother, who was a slave born in 1790. Her name was Judy. That's the only name that we have for her, because of the lack of surnames for slaves at that time.

I've tried to trace all sides of my family, every branch. I didn't want to leave any stones unturned. Perhaps the most notable were the Barton family, the family of my great-great-great-grandparents, Floyd and Ellen Barton, both of whom were slaves in Georgia, on a plantation near a town called Macon, in Bibb County.

Floyd and Ellen gave birth to five children, and one of them, a daughter, is the one that I'm directly descended from. She would be my great-great-grandmother. Her name was Tallulah Barton. She was born in 1865, the year the slaves were emancipated. She married a man named Henry Casterlow. Of course, his last name was taken from the slave owner whose family was from England. In doing my research, I came across a distant cousin of the man who had owned my family as slaves. And he shared information about the family.

Lola Mae Harris, who lived from 1916 to 1968, is Skip Mason's maternal grandmother.

Q: What would you suggest as a starting point for someone who wants to research their family history?

If nothing else, it's important that the oldest person in the family be contacted. And that was one of the most crucial parts of my research, when I contacted my great-grandmother's elder sister, Aunt Minnie. She was the daughter of Henry Casterlow and Tallulah Barton Casterlow and was born in 1888. Lived in Gainesville, Florida, and I had heard about her all of my life. When family members heard that I was interested in researching the family, they would tell me that I had to talk to Aunt Minnie.

One Saturday afternoon in August of 1982, I picked up the phone and called Aunt Minnie, spoke to her for the first time in my life. And she mentioned that the family name had changed. Before that time, I had thought it was Smith.

Deloris Harris, Skip's mother, in her graduation picture from Morris Brown College in Atlanta.

But when she told me the family name was Casterlow, I was able to go to the archives and find the correct census records. And in one day, I was able to go back three generations. I found Henry Casterlow, my great-great-grandfather. I found his father, and I found his grandfather, which took me back to my great-great-great-great-grandfather. His name was Juba Casterlow.

I was just so excited, I could have shouted. I could have told the entire world what I had found. I wanted to call Aunt Minnie back and let her know. But I decided that I would wait until the weekend, because the telephone rates were much cheaper then. Well, that Friday night I called Aunt Minnie's house, and her daughter answered the phone. And she sounded a little like she had been crying. So I asked her, "What's wrong?" And she said, "Mama died," meaning that Aunt Minnie had died. It had been just four days since I spoke with her. But she told me that her Mama was so excited the whole week over the fact that I had started researching the family history, because she had always wanted somebody to do that.

Herman Mason, Sr., Skip's father, in his graduation picture from the class of 1949 at Stanton High School.

Well, after that I became obsessed with recording the fam-

The "sisters," a group of Skip Mason's cousins from Cleveland, Ohio, in the 1940s.

ily history. I couldn't sleep. All I could do was think family history. Making phone calls...I went to the archives literally every day. I sent out letters, I mean I was a basket case over genealogy. I *had to know*. With Aunt Minnie's death, it was imperative, because Aunt Minnie took with her 92 years of history. And to this day, I think of all the questions I could have asked her.

Q: Tell us about your parents.

They were both born in Jacksonville, Florida, and went to Morris Brown College here in Atlanta. That was where they met. Later, my father worked at the Lockheed Company and my mother was a schoolteacher. My mother was the first person in her family to go to college. And that was a big to-do. I understand that my grandmother went up the street for days and days telling the neighbors that her daughter had gone to college. It was a big accomplishment, a proud moment for the family.

But they also say that each generation should get better and better. There was no question whether or not I would go to college. I was just raised, prepared to go to college. I majored in communications, with a minor in history. I thought I needed a career so that I could earn some money. And history didn't seem to be the way, at that time. But I went to Atlanta University and got a master's degree in history.

Then I worked at the Atlanta Public Library for five years. It was one of the greatest things that happened to me. As the black studies archivist, I had my fingers on every source available, every book, microfilmed newspapers. Then, two years ago, in 1992, I had a vision and decided it was time for me to leave the library and start my own company, Digging It Up.

Q: You've been quite successful with it.

Members of the McRae and Tobler families, Skip Mason's cousins, at a family gathering in Lumber City, Georgia, around 1944.

Since I started researching my family history, I began to help others do the same. I decided to create a company to do just that. I have a collection of photographs, and often people will call—documentary producers, writers—looking for pictures of African Americans. Spike Lee, the movie director, contacted me when he was doing a movie called *School Daze* and he wanted to open the picture with a montage of photographs. They used many of my images, and the thing that excited me was that I got paid for it. And I said, "Hey, here's an idea." I decided to create my company. We supply images to people for exhibits, for documentaries, for advertising, for any kind of visual media. We also specialize in African American family research. We have a genealogy packet. We want to make it affordable, for genealogy research can be expensive.

As you know, the Olympics will be held here in Atlanta in 1996. I was selected as the project historian to research, identify, and write text for historical markers dealing with African Americans in Atlanta. And it's been exciting work. Significant sites previously unknown to Atlantans will now be known.

One thing that I have learned. I believe in divine intervention. When it's time for something to happen, it will happen. Things will work out. Sometimes we rush, and try to make things happen before it's time.

Q: I think you've already answered this, but what does it give you to know the history of your family? Why is that important?

Well, first of all it lets you know that the world didn't start with you. It gives you a feeling of continuity, helps you to know who you look like, why you behave this way.

It connects you with history, gives you a sense of time and place, because you can't live isolated in this society. For every historical event that occurred, you ought to be able to identify some relative of yours who was a part of it. I can go and ask my mom, "Where were you when Martin Luther King died?" She was a part of that time and place. I was in the first grade. I knew something was going on, because my mother and I were very close, and the day of the funeral my mother got me out of school to leave me with a neighbor. She told me she was going to march in the funeral procession. I cried because I wanted to march too. I have this recollection of her telling me that Dr. King had died. So every time I look at the films of the procession from the church to the graveyard, I know that my mother was out there with the thousands of people.

My grandfather was a part of World War II, and I had a relative who fought in France during World War I. In 1865, when the slaves were freed, I can say which of my relatives were there, which plantations they lived on.

So knowing your family history allows you to put things into historical perspective. Because doing family history is much more than just gathering names and dates. When you look at a headstone, a tombstone, there are two dates on it—the date you were born, and the date that you died. Those two dates are relatively insignificant. The most important aspect is that little dash in the middle. That dash indicates what you did, what kind of life you had, what kind of work you performed. Everything that revolves around a person is contained in that dash. And that's what's important. What kinds of things you did to make some kind of lasting contribution. I do believe, as I told a friend of mine, that we are here to do something.

Many of my ancestors were farmers and sharecroppers. But it was important in terms of what they did, to feed their families, to provide clothing and shelter. So in finding out about their lives, I become a fuller person. As I tell my students at Morehouse, it's wonderful that you know about Frederick Douglass, Harriet Tubman, Malcolm X, Martin Luther King, and all the other great African Americans. But if you cannot tell me anything about your great-grandfather, then I have problems about that. Because that is the person you descend from, and you need to know more about him than about someone else. Because that person's life is important within the scheme of your life, in terms of your getting here to this particular point.

Deloris Harris Mason is the bridesmaid at the far right of the first row in this wedding in the 1950s.

The cover of Going Against the Wind, *Skip Mason's pictorial history of blacks in Atlanta. In 1994, his company was chosen to create exhibits for a new civil rights museum in Savannah, Georgia.*

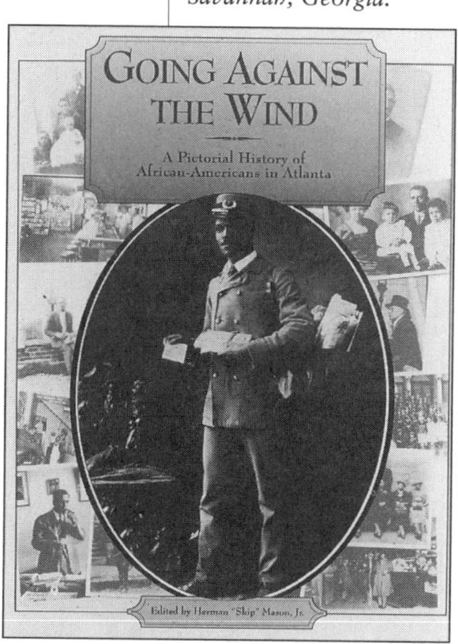

AFRICAN AMERICAN TIMELINE

1526
First Africans arrive, with the Spanish, in what is now South Carolina.

1619
10 Africans arrive in Jamestown, Virginia; they are the first to inhabit a British North American colony.

1770s
Jean Baptiste Point du Sable establishes first settlement at the site of present-day Chicago.

1774
Rhode Island abolishes slavery; seven other northern states pass similar laws by 1789.

1808
Congress bans further importation of slaves into the United States.

1831
Nat Turner leads slave revolt in Virginia.

1850
Fugitive Slave Act makes it a federal crime to assist an escaped slave.

January 1, 1863
Abraham Lincoln issues the Emancipation Proclamation, freeing slaves in states that had seceded from the Union.

December 18, 1865
13th Amendment is ratified, abolishing slavery in the United States.

1868
14th Amendment grants the rights of citizenship to former slaves.

1870
15th Amendment guarantees the right to vote to former male slaves and their male descendants.

1877
Reconstruction era ends with the withdrawal of federal troops from the South.

1879
Benjamin "Pap" Singleton organizes an exodus of African Americans from southern states into Kansas.

1881
Booker T. Washington becomes first head of Tuskegee Institute in Tuskegee, Alabama.

1896
In *Plessy* v. *Ferguson,* the Supreme Court rules that states can maintain "separate but equal" public accommodations for whites and blacks.

1905
W. E. B. Du Bois and other African American leaders found the Niagara Movement to work for equal rights for blacks.

Robert S. Abbott founds the *Chicago Defender,* a newspaper that urges southern blacks to move North. Abbott's editorials help to start the Great Migration.

1909
Founding of the National Organization for the Advancement of Colored People (NAACP)

1926
Historian Carter G. Woodson starts Negro History Week (now Black History Month).

1930
First Temple of Islam founded in Detroit by Wali Farad. Elijah Muhammad would lead the Nation of Islam from 1934 to 1975.

1941
President Franklin D. Roosevelt issues Executive Order 8802, which bars racial discrimination in the defense industry.

1947
Jackie Robinson joins the Brooklyn Dodgers and breaks the racial barrier of baseball's major leagues.

1954
In *Brown* v. *Board of Education,* the Supreme Court requires states to integrate their school systems.

December 1, 1955
In Montgomery, Alabama, Rosa Parks refuses to give up her seat on a bus to a white man and is arrested. The bus boycott that followed was the beginning of the modern civil rights movement.

August 28, 1963
March on Washington to support pending civil rights legislation draws 250,000 participants.

1964
Congress passes first Civil Rights Act since Reconstruction.

1967
Thurgood Marshall becomes the first African American justice of the Supreme Court.

April 4, 1968
Martin Luther King, Jr., is assassinated in Memphis, Tennessee.

1988
Jesse Jackson is the first African American to win state primary elections in his bid to win the Democratic nomination for President of the United States.

FURTHER READING

General Accounts of African American History

Bennett, Lerone, Jr. *Before the Mayflower.* New York: Penguin, 1982.

Bennett, Lerone, Jr. *The Shaping of Black America.* Chicago: Johnson, 1975.

Berlin, Ira, et al., eds. *Free at Last: A Documentary History of Slavery, Freedom, and the Civil War.* New York: New Press, 1992.

Katz, William Loren. *Eyewitness: The Negro in American History.* New York: Pitman, 1971.

Quarles, Benjamin. *The Negro in the Making of America.* 3rd ed. New York: Macmillan, 1987.

Specific Aspects of African American History

Du Bois, W. E. B. *The Souls of Black Folk: Essays and Scketches.* New York: Dodd, Mead, 1979.

Johnson, James Weldon. *Black Manhattan.* New York: Atheneum, 1968.

Lewis, David Levering. *When Harlem Was in Vogue.* New York: Oxford University Press, 1981.

Painter, Nell Irvin. *Exodusters: Black Migration to Kansas after Reconstruction.* New York: Knopf, 1977.

Smith, Jessie Carney. *Epic Lives: One Hundred Black Women Who Made a Difference.* Detroit: Visible Ink Press, 1993.

First-Person Accounts of African American Life

Andrews, William L., ed. *Six Women's Slave Narratives.* New York: Oxford University Press, 1988.

Angelou, Maya. *I Know Why the Caged Bird Sings.* New York: Random House, 1969.

Baldwin, James. *Notes of a Native Son.* Boston: Beacon Press, 1955.

Bates, Daisy. *The Long Shadow of Little Rock: A Memoir.* New York: David McKay, 1962.

Bontemps, Arna, ed. *Great Slave Narratives.* Boston: Beacon Press, 1969.

Brown, Claude. *Manchild in the Promised Land.* New York: Macmillan, 1965.

Hughes, Langston. *The Big Sea.* New York: Knopf, 1940.

Hurston, Zora Neale. *Dust Tracks on a Road.* New York: Harper Perennial, 1991.

Johnson, John H., with Lerone Bennett. *Succeeding against the Odds.* New York: Amistad Press, 1992.

King, Martin Luther, Jr. *I Have a Dream: Writings and Speeches that Changed the World.* Edited by James M. Washington. San Francisco: Harper/San Francisco, 1992.

Malcolm X with Alex Haley. *The Autobiography of Malcolm X.* New York: Grove Press, 1965.

Meltzer, Milton. *The Black Americans: A History in Their Own Words, 1619–1983.* New York: Harper & Row, 1984.

Robinson, Jackie. *Baseball Has Done It.* Edited by Charles Dexter. Philadelphia: Lippincott, 1964.

Ross, Diana. *Secrets of a Sparrow.* New York: Villard, 1993.

Taulbert, Clifton L. *Once upon a Time When We Were Colored.* Tulsa, Okla.: Council Oak Books, 1989.

Washington, Booker T., *Up from Slavery.* New York: Penguin, 1986.

Poems, Novels, and Plays by African Americans

Adoff, Arnold, ed. *The Poetry of Black America.* New York: HarperCollins, 1973.

Baldwin, James. *Go Tell It on the Mountain.* New York: Knopf, 1953.

Baraka, Imamu Amiri. *Selected Poetry of Amiri Baraka–LeRoi Jones.* New York: William Morrow, 1979.

Brooks, Gwendolyn. *Selected Poems.* New York: Harper & Row, 1963.

Ellison, Ralph. *Invisible Man.* New York: Random House, 1972.

Hansberry, Lorraine. *A Raisin in the Sun.* New York: New American Library, 1966.

Hughes, Langston. *Selected Poems of Langston Hughes.* New York: Random House, 1974.

Hurston, Zora Neale. *Their Eyes Were Watching God.* New York: Harper, 1990.

Miller, E. Ethelbert, ed. *In Search of Color Everywhere: A Collection of African-American Poetry.* New York: Stewart, Tabori & Chang, 1994.

Morrison, Toni. *Jazz.* New York: Knopf, 1992.

Parks, Gordon. *The Learning Tree.* New York: Harper & Row, 1963.

Walker, Alice. *The Color Purple.* San Diego: Harcourt Brace Jovanovich, 1982.

Wright, Richard. *Native Son.* New York: Harper & Row, 1940.

TEXT CREDITS

Main Text

p. 12: [Venture], *A Narrative of the Life and Adventures of Venture, a Native of Africa...Related by Himself* (New London, Conn., 1798; revised and republished, Haddam, Conn.: H. M. Selden, 1896), 5-12.

p. 13: Olaudah Equiano, *The Life of Olaudah Equiano, or Gustavus Vassa, the African,* in *Great Slave Narratives,* ed. Arna Bontemps (Boston: Beacon Press, 1969), 4-15.

p. 15: Michael Olatunji, interview with Margo Nash, from files of Statue of Liberty National Monument, November 19, 1973, 1-3.

p. 16: D. T. Niane, *Sundiata: An Epic of Old Mali* (Harlow, Essex, England: Longman, 1965), 13, 15, 19-21, 41, 49-50, 81-82.

p. 20: *A Narrative of the Life and Adventures of Venture,* 5-12.

p. 21, top: Equiano, *The Life of Olaudah Equiano,* 2.

p. 21, bottom: [Louis Asa-Asa], "Narrative of Louis Asa-Asa, a captured African," in *The History of Mary Prince, a West Indian Slave, Related by Herself,* ed. Moira Ferguson (1831; reprint, London: Paulon, 1987), 122-24.

p. 22: Equiano, *The Life of Olaudah Equiano,* 27-28.

p. 23: Equiano, *The Life of Olaudah Equiano,* 30-31.

p. 24, top: Asa-Asa, "Narrative of Louis Asa-Asa," 124.

p. 24, bottom: Equiano, *The Life of Olaudah Equiano,* 31-33.

p. 30: Richard Hakluyt, *The Foreign Voyages* (1600; reprint, London: J.M. Dent, 1928), 155-57.

p. 31, top: Andrés González de Barcia Carballido y Zuñiga, *Barcia's Chronological History of the Continent of Florida* (1722; reprint, Westport, Conn.: Greenwood Press, 1970), 70.

p. 31, middle: Mrs. John Kinzie, *Wau-bun: The "Early Day" in the North-West.* (Chicago: Lakeside Press, 1932 facsimile of 1856 edition), 219-20.

p. 31, bottom: Thomas A. Meehan, "Jean Baptiste Point du Saible, the First Chicagoan," *Mid-America,* April 1937, 87-89.

p. 32: Milton Meltzer, *The Black Americans: A History in Their Own Words, 1619–1983* (New York: Harper & Row, 1984), 23-27.

p. 33: Norman R. Yetman, ed., *Life under the "Peculiar Institution": Selections from the Slave Narrative Collection* (New York: Holt, Rinehart, & Winston, 1970), 69-70.

p. 34: Belinda Hurmence, ed., *Before Freedom: 48 Oral Histories of Former North and South Carolina Slaves* (New York: Mentor Books, 1990), 154-56.

p. 35, top: Bruce Levine et al., *Who Built America?,* vol. 1 (New York: Pantheon, 1989), 212.

p. 35, bottom: Elizabeth Keckley, *Behind the Scenes* (New York: Oxford University Press, 1988), 34.

p. 36, top: Henry Louis Gates, Jr., ed., *The Classic Slave Narratives* (New York: Mentor Books, 1987), 269-70.

p. 36, middle: E. Franklin Frazier, *The Negro Family in the United States* (Chicago: University of Chicago Press, 1939), 48.

p. 36, bottom: From *I Wish I Could Give My Son a Wild Raccoon* by Eliot Wigginton. Copyright © by Reading is Fundamental. Used by permission of Doubleday, a division of Bantam Doubleday Dell Publishing Group, Inc., pp. 268-69.

p. 37: Virginia Writer's Project. *The Negro in Virginia* (New York: Arno Press, 1969), 81.

p. 38, top: *Memoir of Old Elizabeth, a Coloured Woman* (New York: Oxford University Press, 1988), 3-4.

p. 38, bottom: Frazier, *The Negro Family,* 42.

p. 39, top: James Mellon, ed., *Bullwhip Days: The Slaves Remember* (New York: Avon Books, 1990), 121.

p. 39, middle: James Williams, *Narrative of James Williams* (Boston: Isaac Knapp, 1838; Philadelphia: Rhistoric Publications, 1969), 28-29.

p. 39, bottom: William Loren Katz, *Eyewitness: The Negro in American History* (New York: Pitman, 1971), 133.

p. 40, top: George Ducas and Charles Van Doren, eds., *Great Documents in Black American History* (New York: Praeger, 1970), 24-25.

p. 40, bottom: Margaret Busby, ed., *Daughters of Africa* (New York: Ballantine, 1992), 43.

p. 41: Levine, *Who Built America?,* 208.

p. 42, top: Levine, *Who Built America?,* 379.

p. 42, bottom: Charlotte Forten Grimké, *The Journals of Charlotte Forten Grimké,* ed. Brenda Stevenson (New York: Oxford University Press, 1988), 139-40.

p. 43, top: Giles Wright, ed., *Looking Back: Eleven Life Histories* (Trenton, N.J.: New Jersey Historical Commission, 1986), 58.

p. 43, bottom: Frazier, *The Negro Family,* 135.

p. 44: Herbert Aptheker, *Nat Turner's Slave Rebellion* (New York: Humanities Press, 1966), 134-38.

p. 45: Williams, *Narrative of James Williams,* 83-86.

p. 46: George Stearns, *Narrative of Henry Box Brown by Himself* (Boston: Brown and Stearns, 1849), 60-62.

p. 47: Mattie J. Jackson, *The Story of Mattie J. Jackson,* written and arranged by L. S. Thompson, in *Six Women's Slave Narratives* (New York: Oxford University Press, 1988), 26-28.

p. 48, top: Carter G. Woodson, ed., *The Mind of the Negro as Reflected in Letters Written during the Crisis, 1800–1860* (New York: Negro Universities Press, 1969), 544.

p. 48, bottom: Hurmence, *Before Freedom,* 12-14.

p. 49, top: Levine, *Who Built America?,* 451.

p. 49, bottom: Frazier, *The Negro Family,* 75-76.

p. 54, top: Hurmence, *Before Freedom,* 32-33.

p. 54, bottom: Frazier, *The Negro Family,* 132-33.

p. 55: Hamilton Holt, ed., *The Life Stories of Undistinguished Americans, As Told by Themselves* (New York: Routledge, 1990), 114-15.

p. 56: Lerone Bennett, Jr., *The Shaping of Black America* (Chicago: Johnson, 1975), 63.

p. 57: Holt, *Life Stories,* 119-23.

p. 58: Katz, *Eyewitness: The Negro in American History,* 81-82.

p. 59: Nell Irvin Painter, *Exodusters: Black Migration to Kansas after Reconstruction* (New York: Knopf, 1977), 3-4.

p. 60, top: From *I Wish I Could Give My Son a Wild Raccoon* by Eliot Wigginton. Copyright © by Reading is Fundamental. Used by permission of Doubleday, a division of Bantam Doubleday Dell Publishing Group, Inc., p. 269.

p. 60, bottom: Nat Love, *The Life and Adventures of Nat Love, by Himself* (New York: Arno Press, 1968), 73

p. 61: Era Bell Thompson, *American Daughter* (St. Paul: Minnesota Historical Society Press, 1986), 73-74.

p. 62, top: Levine, *Who Built America?,* 506.

p. 62, bottom: Hurmence, *Before Freedom,* 84.

p. 63: Holt, *Life Stories,* 220-21.

p. 64, top: From *I Wish I Could Give My Son a Wild Raccoon* by Eliot Wigginton. Copyright © by Reading is Fundamental. Used by permission of Doubleday, a division of Bantam Doubleday Dell Publishing Group, Inc., p. 274.

p. 64, bottom: Levine, *Who Built America?,* 155.

p. 65: Walter White, *A Man Called White* (New York: Viking, 1948), 9-12.

p. 66, top: Emmett J. Scott, ed., "Letters of Negro Migrants of 1916–1918," *Journal of Negro History* 4 (July 1919, October 1919): 17-18, 290, 346.

p. 66, bottom: Giles R. Wright and Howard L. Green, *Work,* New Jersey Ethnic Life Series (Trenton: New Jersey Historical Commission, 1987), 16.

p. 67: Giles R. Wright, *The Journey from Home,* New Jersey Ethnic Life Series (Trenton: New Jersey Historical Commission, 1986), 19-20.

p. 68: John H. Johnson, with Lerone Bennett, Jr., *Succeeding against the Odds* (New York: Amistad Press, 1992), 47, 51, 58-59.

p. 69: Jackie Robinson, *Baseball Has Done It,* ed. Charles Dexter (Philadelphia: Lippincott, 1964), 27-29.

p. 74, top: Richard Meryman, interviewer, *The Life and Thoughts of Louis Armstrong—A Self-portrait* (New York: Eakins Press, 1971).

p. 74, bottom: James P. Comer, *Maggie's American Dream* (New York: New American Library, 1988), 34-35.

p. 75, top: Giles R. Wright, *Arrival and Settlement in a New Place,* New Jersey Ethnic Life Series (Trenton: New Jersey Historical Commission, 1986), 14.

p. 75, bottom: Wright, *Journey from Home,* 37-38.

p. 76: From *Refuse to Stand Silently By* by Eliot Wigginton. Copyright © 1992 by Highlander Center. Used by permission of Doubleday, a division of Bantam Doubleday Dell Publishing Group, Inc., pp. 185-86.

p. 77: Thomas C. Wheeler, ed., *The Immigrant Experience: The Anguish of Becoming American* (New York: Penguin Books, 1992), 133-34.

p. 78, top: Wright and Green, *Work,* 31-32.

p. 78, bottom: Scott, "Letters of Negro Migrants," 413.

p. 79: Robert Asher, "Documents of the Race Riot at East St. Louis," *Illinois State Historical Society Journal* 65, no. 3 (Autumn 1972), 328-29.

p. 80, top: Langston Hughes, *The Big Sea* (New York: Hill & Wang, 1940), 32-33.

p. 80, bottom: Robinson, *Baseball Has Done It,* 29-32.

p. 81, top: Daisy Bates, *The Long Shadow of Little Rock: A Memoir* (New York: David McKay, 1962), 6-7.

p. 81, bottom: Zora Neale Hurston, *Dust Tracks on a Road* (New York: Harper Perennial, 1991), 11-12.

p. 82, top: Excerpts from *You Must Remember This: An Oral History of Manhattan from the 1890s to World War II,* copyright © 1989 by Jeff Kisseloff, reprinted by permission of Harcourt Brace & Company, p. 265.

p. 82, bottom: Lofton Mitchell, "Harlem Reconsidered," *Freedomways,* Fall 1964, 469-70.

p. 84, top: Gordon Parks, *Voices in the Mirror: An Autobiography* (New York: Doubleday, 1990), 6-7.

p. 84, bottom: From *I Wish I Could Give My Son a Wild Raccoon* by Eliot Wigginton. Copyright © by Reading is Fundamental. Used by permission of Doubleday, a division of Bantam Doubleday Dell Publishing Group, Inc., pp. 166-67.

p. 85: From *Refuse to Stand Silently By* by Eliot Wigginton. Copyright © 1992 by Highlander Center. Used by permission of Doubleday, a division of Bantam Doubleday Dell Publishing Group, Inc., pp. 172-73.

p. 86: Bates, *The Long Shadow,* 7-9.

p. 88, top: James Baldwin, *Notes of a Native Son* (Boston: Beacon Press, 1955), 86-88.

p. 88, bottom: Excerpt from *Once upon a Time When We Were Colored* © 1989 by Clifton L. Taulbert, published by Council Oak Books, Tulsa, Oklahoma, pp. 10-11.

p. 89: Wheeler, *The Immigrant Experience,* 137-38.

p. 90: Comer, *Maggie's American Dream,* 43-44.

p. 91, top: Malcolm X with Alex Haley, *The Autobiography of Malcolm X* (New York: Grove Press, 1966), 194-95.

p. 91, bottom: Excerpt from *Once upon a Time When We Were Colored* © 1989 by Clifton L. Taulbert, published by Council Oak Books, Tulsa, Oklahoma, pp. 91-95.

p. 94: Mary McLeod Bethune, "Faith That Moved a Dump Heap," *Who, the Magazine about People,* June 1941, 31-35.

p. 95, top: Giles R. Wright, *Schooling and Education,* New Jersey Ethnic Life Series (Trenton: New Jersey Historical Commission, 1987), 26.

p. 95, bottom: From *Refuse to Stand Silently By* by Eliot Wigginton. Copyright © 1992 by Highlander Center. Used by permission of Doubleday, a division of Bantam Doubleday Dell Publishing Group, Inc., pp. 279-80.

p. 96, top: Jake Lamar, *Bourgeois Blues* (New York: Summit Boooks, 1991), 17, 20-21.

p. 96, bottom: Wright, *Schooling and Education,* 29.

p. 97: Hughes, *The Big Sea,* 27-31.

p. 102, top: Joshua Freeman et al., *Who Built America?,* vol. 2 (New York: Pantheon Books, 1992), 540-41.

p. 102, bottom: Freeman, *Who Built America?,* 539-40.

p. 103: Bates, *The Long Shadow,* 73-76.

p. 104: Freeman, *Who Built America?,* 547.

p. 105: George Breitman, ed., *Malcolm X Speaks* (New York: Grove Press, 1965), 138-45.

p. 106: Used by permission of The University of Alabama Press from *Selma, Lord Selma: Girlhood Memories of the Civil-Rights Days,* by Sheyann Webb and Rachel West Nelson, as told to Frank Sikora. © 1980 The University of Alabama Press, pp. 95-98.

p. 107: From *Voices of Freedom* by Henry Hampton and Steve Fayer. Copyright © 1990 by Blackslide, Inc. Used by permission of Bantam Books, a division of Bantam Doubleday Dell Publishing Group, Inc., pp. 389-90.

p. 108: From *Voices of Freedom* by Henry Hampton and Steve Fayer. Copyright © 1990 by Blackslide, Inc. Used by permission of Bantam Books, a division of Bantam Doubleday Dell Publishing Group, Inc., pp. 467-68.

p. 109, top: From *Voices of Freedom* by Henry Hampton and Steve Fayer. Copyright © 1990 by Blackslide, Inc. Used by permission of Bantam Books, a division of Bantam Doubleday Dell Publishing Group, Inc., pp. 640-41.

p. 109, bottom: From *I Wish I Could Give My Son a Wild Raccoon* by Eliot Wigginton. Copyright © by Reading is Fundamental. Used by permission of Doubleday, a division of Bantam Doubleday Dell Publishing Group, Inc., p. 169.

p. 110, top: Excerpt from *Once upon a Time When We Were Colored* © 1989 by Clifton L. Taulbert, published by Council Oak Books, Tulsa, Oklahoma, pp. 51-52.

p. 110, bottom: Jackie Robinson, as told to Wendell Smith, *Jackie Robinson: My Own Story* (New York: Greenberg, 1948), 128.

p. 111, top: Hank Aaron with Lonie Wheeler, *I Had a Hammer* (New York: Harper Collins, 1991), 268-70.

p. 111, bottom: Kareem Abdul-Jabbar with Mignon McCarthy, *Kareem* (New York: Random House, 1990), 9-11.

p. 112, top: Muddy Waters, interview with Jan O'Neal and Amy O'Neal, courtesy of Delta Blues Museum, Clarksdale, Mississippi.

p. 112, bottom: Johnson, *Succeeding against the Odds,* 2-4.

p. 113: Terry McMillan et al., *The Films of Spike Lee* (New York: Stewart, Tabori & Chang, 1991), iv-v.

p. 114, top: Excerpts from *You Must Remember This: An Oral History of Manhattan from the 1890s to World War II,* copyright © 1989 by Jeff Kisseloff, reprinted by permission of Harcourt Brace & Company, pp. 305-6.

p. 114, bottom: Excerpts from *You Must Remember This: An Oral History of Manhattan from the 1890s to World War II,* copyright © 1989 by Jeff Kisseloff, reprinted by permission of Harcourt Brace & Company, p. 303.

p. 115, top: Excerpts from *You Must Remember This: An Oral History of Manhattan from the 1890s to World War II,* copyright © 1989 by Jeff Kisseloff, reprinted by permission of Harcourt Brace & Company, p. 313.

p. 115, bottom: Diana Ross, *Secrets of a Sparrow* (New York: Villard, 1993), 91-97.

p. 116, top: Eric V. Copage, *Kwanzaa: An African-American Celebration of Culture and Cooking.* (New York: William Morrow, 1991), xiii-xiv.

p. 116, bottom: Studs Terkel, *Race: How Blacks and Whites Think and Feel about the American Obsession* (New York: Anchor Books, 1993), 170.

Sidebars

p. 13: Bruce Levine et al., *Who Built America?,* vol. 1 (New York: Pantheon, 1989), 23.

p. 15: "Afro-American Fragment," from *Selected Poems,* by Langston Hughes, copyright © 1959 by Langston Hughes, reprinted by permission of Alfred A. Knopf, Inc.

p. 17: Sir Richard F. Burton, comp., *Wit and Wisdom from West Africa* (1865; reprint, New York: Biblio & Tannen, 1969), 5.

p. 20: Levine, *Who Built America?,* 26.

p. 21: Phillis Wheatley, *Poems on Various Subjects* (London, 1773), 74.

p. 24: Theodore Canot, *Adventures of an African Slaver,* ed. Malcolm Cowley (1854; reprint, Garden City, N.Y.: Doubleday, 1928), 107-8, 110-11.

p. 32: Herbert G. Gutman archive, American Social History Project, City University of New York.

p. 34: Frederick Law Olmsted, *A Journey in the Seaboard Slave States in the Years 1853–1854 with Remarks on Their Economy,* vol. 2 (New York: Putnam, 1904), 218-19.

p. 35: Cecyle S. Neidle, *The New Americans* (New York: Twayne, 1967), 97.

p. 38: William Loren Katz, *Eyewitness: The Negro in American History* (New York: Pitman, 1971), 104.

p. 39: Frederick Douglass, *The Life and Times of Frederick Douglass* (1892; reprint, New York: Macmillan, 1962), 146, 147.

p. 40: "Reminiscences by Frances D. Gage," in Elizabeth Cady Stanton, Susan B. Anthony, and Matilde Joselyn Gage, eds., *History of Woman Suffrage* (1881; reprint, New York: Source Book Press, 1970), 115-17.

p. 42: Alexis de Tocqueville, *Democracy in the United States,* vol. 1 (New York: New American Library, 1945), 373-74.

p. 43: Leslie H. Fishel, Jr., and Benjamin Quarles, eds. *The Negro American: A Documentary History* (Glenview, Ill.: Scott, Foresman, 1967), 156.

P. 46: Carter G. Woodson, ed., *The Mind of the Negro as Reflected in Letters Written during the Crisis: 1800-1860* (New York: Negro Universities Press, 1969), 508-9.

p. 49: Herbert Aptheker, *A Documentary History of the Negro People of the United States,* vol. 1 (New York: Citadel Press, 1990), 460-61.

p. 56: Joshua Freeman et al., *Who Built America?,* vol. 2 (New York: Pantheon Books, 1992), 45.

p. 59: Nell Irvin Painter, *Exodusters: Black Migration to Kansas after Reconstruction* (New York: Knopf, 1977), 114.

p. 67: Clifton L. Taulbert, *The Last Train North* (Tulsa, Okla.: Council Oak Books, 1992), 17-21.

p. 68: Allan H. Spear, *Black Chicago: The Making of a Negro Ghetto, 1890–1920* (Chicago: University of Chicago Press, 1967), 137.

p. 69: Howard L. Green and Lee R. Parks, *What Is Ethnicity?* (Trenton: New Jersey Historical Commission, 1987), 25.

p. 78: Excerpts from *You Must Remember This: An Oral History of Manhattan from the 1890s to World War II,* copyright © 1989 by Jeff Kisseloff, reprinted by permission of Harcourt Brace & Company, p. 298.

p. 80: Richard Wright and Edwin Rosskam, *Twelve Million Black Voices* (1941; reprint, New York: Thunder's Mouth Press, 1988), 104-5.

p. 82: Katz, *Eyewitness,* 393.

p. 83: Freeman, *Who Built America?,* 306.

p. 89: Harriet Pipes McAdoo, ed., *Black Families* (Newbury Park, Calif.: Sage, 1988), 291.

p. 91: George A. Singleton, ed., *The Life Experience and Gospel Labors of the Rt. Rev. Richard Allen* (New York: Abingdon Press, 1960), 25-26.

p. 97: Studs Terkel, *Race: How Blacks and Whites Think and Feel about the American Obsession* (New York: Anchor Books, 1992), 182.

p. 114: Excerpts from *You Must Remember This: An Oral History of Manhattan from the 1890s to World War II,* copyright © 1989 by Jeff Kisseloff, reprinted by permission of Harcourt Brace & Company, pp. 317-18.

p. 116: Vertamae Smart-Grosvenor, *Vibration Cooking: Or the Travel Notes of a Geechee Girl* (New York: Ballantine, 1992), 21-23.

PICTURE CREDITS

Abby Aldrich Rockefeller Folk Art Center, Williamsburg, Va.: 33 bottom; Courtesy the African American Museum and Library at Oakland: cover, 52, 73, 89; courtesy Department of Library Services, American Museum of Natural History: 8 (neg. #325337), 10 (neg. #14087), 11 (neg. #276937); Amistad Research Center: 70, 94 top; the Archives of Labor and Urban Affairs, Wayne State University: 100; Bettmann Archive: 85; Boston Athenaeum: 47 bottom; British Library: 14 top; Chicago Bulls: 110 bottom; Chicago Historical Society: 31 (ICHi-12166, engraved by Raoul Varin), 69 (Raymond Trowbridge, Box 200, #22); Children's Defense Fund: 88; Cook Collection, Valentine Museum: 26, 37, 50, 53, 54; Billy Deal: 111 bottom; Digging It Up: 101, 107 bottom, 118 top and bottom, 119 top and bottom, 120 top and bottom, 121 top and bottom; courtesy George Eastman House: 74 bottom; Freedmen and Southern Society Project, University of Maryland: 48 top and bottom; Idaho State Historical Society: 61 top (#78-203.20), 86 top (#80-83.4/c); Impact Visuals: 25 (Andrew Lichtenstein), 116 (Catherine Smith), 117 top (Deborah G. Moses-Sanks), 117 bottom (Brian Palmer); Institute of Texan Cultures, El Paso: contents page, 30; Jane Addams Memorial Collection, Special Collections, University Library, the University of Illinois at Chicago: 80 top; Kansas Collection, University of Kansas Libraries: 58 top, 75 (Joseph J. Pennell Collection), 90, 95 top; Kansas State Historical Society, Topeka: 59, 94 bottom, 96 top, 97 right; Kate Kunz, courtesy of Princeton University: 115; League of Women Voters of Chicago Records, Special Collections, University Library, the University of Illinois at Chicago: 79 top; Library of Congress: 22 top, 23 bottom, 24, 43, 55 top, 57 top, 66, 68 top, 76, 87 top, 93 bottom left, 93 right; Musée de l'Homme, Paris: 18; courtesy of the National Afro-American Museum and Cultural Center, Wilberforce, Ohio: 16, 86 bottom, 92; National Archives: 32, 41 bottom, 47 top, 55 bottom, 74 top, 82, 87 bottom, 96 bottom, 102, 104, 112 bottom; Collection of the New-York Historical Society: 23 top, 29, 33 top, 34, 42, 49, 62 top, 83; Ohlinger: 113 top and bottom; Merrill Roberts, Jr.: 98; Reuters/Bettmann: 111 top; Sacramento Archives and Museum Collection: frontispiece; Photographs and Print Division, Schomburg Center for Research in Black Culture, the New York Public Library, Astor, Lenox and Tilden Foundations: 12 bottom, 14 bottom, 15, 20, 22 bottom, 28, 35, 36 top, 36 bottom, 38, 40, 41 top, 45, 46, 57 bottom, 60, 62 bottom, 63 top, 68 bottom, 103 top, 105, 106 bottom, 107 top; Security Pacific National Bank Collection/Los Angeles Public Library: 67, 78, 79 bottom, 80 bottom, 84, 93 middle left, 112 top; Smithsonian Institution: 21, 44; U.S. Army Military History Institute, G. B. Pritchard Coll.: 61 bottom; UPI/Bettmann: 12 top, 63 bottom, 93 top, 95 bottom, 103 bottom, 106 top, 108, 109, 110 top; Washington County Historical Society, Fayetteville, Arkansas: 97 left; Western History Collections, University of Oklahoma Library: 58 bottom; the Western Reserve Historical Society, Cleveland, Ohio: 72, 77.

INDEX

References to illustrations are indicated by page numbers in *italics*.

ACKNOWLEDGMENTS

We want to acknowledge our debt to Carolyn Kozo Cole of the Los Angeles Public Library; Robert Haynes of the African American Museum and Library at Oakland; Isabel Jasper of the National Afro-American Museum and Cultural Center; Teresa Roane of the Valentine Museum; Merrill Roberts, Jr.; Leslie S. Rowland of the Freedmen and Southern Society Project of the University of Maryland; and Mary F. Yearwood, Jim Huffman, and Alice Adamczyk of the Schomburg Center for Research in Black Culture. This book would not have been possible without their generosity.

We would also like to express our appreciation for the valuable contributions and assistance we received from Mary Ann Bamberger and Patricia Bakunas of the Library of the University of Illinois at Chicago; Samuel W. Black and Ann Sindelar of the Western Reserve Historical Society; Diane Bruce of the Institute of Texan Cultures; Tara Deal and Nancy Toff, our editors at Oxford University Press; Dennis D. Dickerson, historiographer of the A.M.E. Church; Meredith Earley of the Delta Blues Museum; Tom Featherstone of the Walter P. Reuther Library; Eileen Flanagan of the Chicago Historical Society; Jim Francis of the New-York Historical Society; Debbie Goodsite of the Bettmann Archive; Elizabeth P. Jacox of the Idaho State Historical Society; Faye Jonason of the California Afro-American Museum; Thomas Jordan of the Washington County Historical Society; Kathy Lafferty of the University of Kansas Libraries; John Lovett of the University of Oklahoma Library; Janice Madhu of the International Museum of Photography; Carol Miguelino of the Lyon County Historical Museum; Charlene Noyes of the City of Sacramento History and Science Division; Lizann Pilgrim of the Marriott Library at the University of Utah; Tony Pisani of the Museum of the City of New York; Philippe Revol of the Musée de l'Homme; Paul Sigrist of the Ellis Island Museum; Catharina Slautterback of the Boston Athenaeum; Brenda B. Square of the Amistad Research Center; Christie Stanley of the Kansas State Historical Society; Zima Thomas of the Children's Defense Fund; Dale Treleven of the UCLA Oral History Program; and Michael J. Winey of the U.S. Army Military History Institute.

Finally, we wish to thank Skip Mason for opening his own African American family album to us and for taking time out from his busy schedule to be interviewed. We regret that this book has space for only a part of his eloquent recollections of his family history and his experiences researching his roots. We envy the young people who are his students at Morehouse College.

ABOUT THE AUTHORS

Dorothy and Thomas Hoobler have published more than 50 books for children and young adults, including *African Portraits; Mandela: The Man, The Struggle, The Triumph; Margaret Mead: A Life in Science; Vietnam: Why We Fought; Showa: The Age of Hirohito;* and *Photographing History: The Career of Mathew Brady.* Their works have been honored by the Society for School Librarians International, the Library of Congress, the New York Public Library, the National Council for Social Studies, and *Best Books for Children,* among other organizations and publications. The Hooblers have also written several volumes of historical fiction for children, including *Next Stop Freedom, Frontier Diary, The Summer of Dreams,* and *Treasure in the Stream.* Dorothy Hoobler received her master's degree in American history from New York University and worked as a textbook editor before becoming a full-time freelance editor and writer. Thomas Hoobler received his master's degree in education from Xavier University, and he previously worked as a teacher and textbook editor.